Contents

If you are following one of the modular specifications, the numbers in brackets indicate which unit(s) the content of each page is relevant to.

D0346428

Place Value

Place Value in Whole Numbers

The value of a digit in a whole number depends on its position in the number. This is called its place value (see the table below). The table below can easily be extended to include ten thousands, hundred thousands, millions etc. This means that it does not matter how large a number is, each digit will still have a place value.

Place Value of Digits				Number
1000 Thousands	**100** Hundreds	**10** Tens	**1** Units	
			7	Seven
		4	2	Forty two
	6	5	9	Six hundred and fifty nine
3	1	8	5	Three thousand, one hundred and eighty five

Rounding Numbers

Numbers can be very large and often make more sense if they are rounded to a given power of 10, e.g. 10 (10^1), 100 (10^2), 1000 (10^3). Let's take the attendance at a football match. The actual attendance is 38 726. However to a neutral observer an attendance figure to the nearest thousand would have more meaning. Look at the place value table below which shows this attendance.

The digit to the right of the one you are working to will tell you if you need to round up. For example, when rounding a number to the nearest thousand you need to look at the digit in the hundreds column. If it is 5 or more, the digit in the thousands column is **rounded up**. If it is 4 or less the digit in the thousands column **stays the same**. The digits in the hundreds, tens and units columns become zero, showing the number to the nearest 1000.

10 000 Ten Thousands	**1 000** Thousands	**100** Hundreds	**10** Tens	**1** Units
3	9	0	0	0
3	8	7	0	0
3	8	7	3	0
3	8	7	2	6

Nearest 1000: Round up

Nearest 100: Stays the same

Nearest 10: Round up

Rounding Numbers

Rounding a Number to 1 (or more) Decimal Place(s)

To round a number to 1 decimal place (1 d.p.) we must look at the **second** digit after the decimal point. There are two possibilities:

1 If the second digit after the decimal point is **4 or less** (i.e. 0, 1, 2, 3 or 4) we leave the first digit after the decimal point as it is.

2 If the second digit after the decimal point is **5 or more** (i.e. 5, 6, 7, 8 or 9) we **round up** and add 1 to the first digit after the decimal point.

To round a number to 2 d.p. we are only allowed two digits after the decimal point. This time we must look at the **third** digit after the decimal point to decide whether we need to round up or to keep the second digit after the decimal point the same. To round a number to 3 d.p. we must look at the **fourth** digit and so on. Unless a question tells you otherwise, always give money to 2 decimal places. Amounts of money are rounded to 2 d.p. in exactly the same way as below.

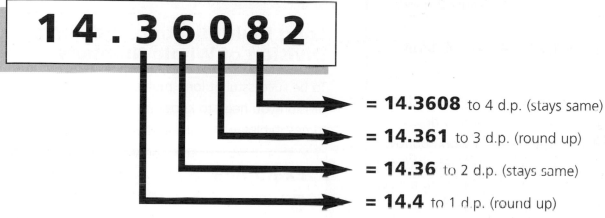

1 4 . 3 6 0 8 2

= **14.3608** to 4 d.p. (stays same)

= **14.361** to 3 d.p. (round up)

= **14.36** to 2 d.p. (stays same)

= **14.4** to 1 d.p. (round up)

Rounding a Number to 1 (or more) Significant Figure(s)

Rounding a number to a certain number of significant figures (s.f.) is very like rounding a number to a certain number of decimal places.

The number alongside is the attendance at a pop concert. It has 4 significant figures.

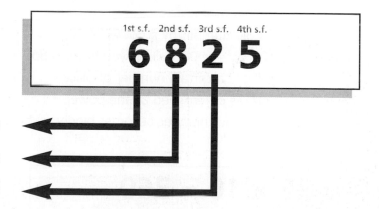

1st s.f.   2nd s.f.   3rd s.f.   4th s.f.

6 8 2 5

7000 to 1 s.f. (round up)

6800 to 2 s.f. (stays same)

6830 to 3 s.f. (round up)

Another Example...

To round numbers less than 1 we follow the same rules except we start counting our significant figures from the first digit greater than 0 (zero).

1st s.f.   2nd s.f.   3rd s.f.   4th s.f.

0 . 0 3 6 1 7

to 1 s.f. is **0.04** (round up)

to 2 s.f. is **0.036** (stays same)

to 3 s.f. is **0.0362** (round up)

Numbers 1

Addition and Subtraction of Whole Numbers

Whenever you add or subtract whole numbers you must line up the digits, one on top of the other, in place value order.

Examples...

1 356 + 72

$$
\begin{array}{r}
356 \\
+\ {}_1 72 \\
\hline
428
\end{array}
$$

- Start from the right hand side.
- **6 + 2 = 8**. Write **8** down.
- **5 + 7 = 12**. Write **2** down and carry **1**.
- **3 + 1 = 4**. Write **4** down.

2 438 – 57

$$
\begin{array}{r}
{}^3 4 {}^1 38 \\
-\quad 57 \\
\hline
381
\end{array}
$$

- Start from the right hand side.
- **8 – 7 = 1**. Write **1** down.
- **3 – 5** doesn't work. Borrow 1 from the hundreds column to give **13 – 5 = 8**. Write **8** down.
- **3 – 0 = 3**. Write **3** down.

Multiplication and Division of Whole Numbers by Powers of 10

To multiply a whole number by a power of 10, e.g. 10(10^1), 100(10^2), 1000(10^3), all you have to do is move all the digits a certain number of place values to the left, as determined by the power. When you do this your number becomes bigger, for example...

1 $36 \times 10 = 360$

2 $36 \times 100 = 3600$

3 $36 \times 1000 = 36\,000$

To divide a whole number by a power of 10 you move all the digits a certain number of places to the right, as determined by the power. When you do this your number becomes smaller, for example...

1 $36 \div 10 = 3.6$

2 $36 \div 100 = 0.36$

3 $36 \div 1000 = 0.036$

Long Multiplication and Long Division of Whole Numbers

To be successful at long multiplication and long division you need to know the multiplication or 'times' tables.

Examples...

1 364 x 14

$$
\begin{array}{r}
364 \\
\times\ 14 \\
\hline
3640 \\
1 4 {}^2 5 {}^1 6 \\
\hline
5096
\end{array}
$$

- **14 = 10 + 4**.
- Do the **364 x 10** first. Remember to put a '**0**' down as you would if you multiplied any whole number by **10**.
- Do the **364 x 4** multiplication.
- Add the two multiplications together.

2 312 ÷ 12

$$
\begin{array}{r}
26 \\
12\overline{)312} \\
24\downarrow \\
\hline
72 \\
72 \\
\hline
0
\end{array}
$$

- **12** does not divide into **3** so move on.
- **12** into **31** goes **2** times. **12 x 2 = 24**. Write **24** below **31** and subtract to give **7**.
- Bring down the **2**. **12** into **72** goes **6** times. **12 x 6 = 72**.

Types of Number

Numbers can be described in many ways. Below is a summary of the types of number that you should know.

Even Numbers

These are numbers which can be divided exactly by 2. The first ten even numbers in order are…

2, 4, 6, 8, 10, 12, 14, 16, 18, 20

Odd Numbers

Since all whole numbers are either even or odd, then odd numbers are those that cannot be divided exactly by 2.

The first ten odd numbers in order are…

1, 3, 5, 7, 9, 11, 13, 15, 17, 19

Factors (Divisors)

The factors (or divisors) of a number are those whole numbers which divide exactly into it. All numbers, with the exception of square numbers (see page 14), have an even number of factors. An easy way to find the factors of a number is to choose pairs of numbers that multiply to give that number. For example

> The factors of **10** are **1, 2, 5, 10**
> (since 1 x 10 = 10, 2 x 5 = 10)
>
> The factors of **24** are **1, 2, 3, 4, 6, 8, 12, 24**
> (since 1 x 24 = 24, 2 x 12 = 24, 3 x 8 = 24, 4 x 6 = 24)

As we said, square numbers have an odd number of factors, for example…

> The factors of **16** are **1, 2, 4, 8, 16**
> (since 1 x 16 = 16, 2 x 8 = 16, 4 x 4 = 16)
>
> The factors of **36** are **1, 2, 3, 4, 6, 9, 12, 18, 36**
> (since 1 x 36 = 36, 2 x 18 = 36, 3 x 12 = 36, 4 x 9 = 36, 6 x 6 = 36)

Multiples

The multiples of a number are those numbers which can be divided exactly by it. To put it simply, they are the numbers found in the 'times' tables.

The multiples of **5** are **5, 10, 15, 20, 25,** etc.

The multiples of **8** are **8, 16, 24, 32, 40,** etc.

Prime Numbers

These are numbers which have only two factors: **1** and **the number itself.** The first ten prime numbers are…

2, 3, 5, 7, 11, 13, 17, 19, 23, 29

The only even prime number is 2 since all even numbers after this have 2 as a factor, which rules them out as prime numbers.

Reciprocals

The reciprocal of a number is '1 over that number', for example…

> **1** the reciprocal of 4 is '1 over 4' $= \frac{1}{4}$
>
> **2** the reciprocal of 0.2 is '1 over 0.2'
> $$= \frac{1}{0.2} \overset{\times 10}{\underset{\times 10}{=}} \frac{10}{2} = 5$$
>
> **3** the reciprocal of $\frac{2}{3}$ is '1 over $\frac{2}{3}$'
> $$= \frac{1}{\frac{2}{3}} = 1 \div \frac{2}{3} = 1 \times \frac{3}{2} = \frac{3}{2} = 1\frac{1}{2}$$

Any non-zero number multiplied by its reciprocal is always equal to 1, for example…

> $$= 4 \times \frac{1}{4} = 1, \ 0.2 \times 5 = 1, \ \frac{2}{3} \times \frac{3}{2} = 1$$

Also, zero has no reciprocal because anything divided by zero is undefined.

Numbers 3

Prime Factor Form

The prime factors of a number are those prime numbers which divide exactly into it. When a number is expressed as a product of its prime factors it is said to be written in prime factor form. To find the prime factor form, try to divide your number by the lowest prime number (i.e. 2). If it works, keep repeating until it will not divide exactly. Then try the next prime number up, and continue until you have an answer of 1. This process is called prime number decomposition.

Examples...

1 Write **24** in prime factor form.

$$\begin{array}{c|c} 2 & 24 \\ 2 & 12 \\ 2 & 6 \\ 3 & 3 \\ & 1 \end{array}$$

so **24 = 2 x 2 x 2 x 3 = 2^3 x 3**

2 Write **420** in prime factor form.

$$\begin{array}{c|c} 2 & 420 \\ 2 & 210 \\ 3 & 105 \\ 5 & 35 \\ 7 & 7 \\ & 1 \end{array}$$

so **420 = 2 x 2 x 3 x 5 x 7 = 2^2 x 3 x 5 x 7**

Alternatively a prime factor tree can be used to work out the prime factors of a number. Take 24 as our example again:

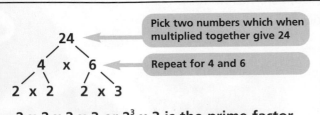

Pick two numbers which when multiplied together give 24

Repeat for 4 and 6

2 x 2 x 2 x 3 or 2^3 x 3 is the prime factor form of 24

Highest Common Factor

The highest common factor (HCF) of two (or more) numbers is the highest factor that divides exactly into both (or all of) the numbers. To find the HCF express your numbers in prime factor form and then select only the prime factors that are common to both numbers.

Example...

What is the HCF of **24** and **90**?

Firstly express **24** and **90** in prime factor form...

24 = 2 x 2 x $\boxed{\text{2 x 3}}$
90 = $\boxed{\text{2 x 3}}$ x 3 x 5

... and then select prime factors that are common to both numbers.

HCF of 24 and 90 = 2 x 3 = 6

Lowest (Least) Common Multiple

The lowest (least) common multiple (LCM) of two or more numbers is the lowest number that is a multiple of all the numbers. You can use prime number decomposition to find the LCM.

Example...

What is the LCM of **8** and **10**?

Write the two numbers down in prime factor form...

$$\begin{array}{c|c} 2 & 8 \\ 2 & 4 \\ 2 & 2 \\ & 1 \end{array} \qquad \begin{array}{c|c} 2 & 10 \\ 5 & 5 \\ & 1 \end{array}$$

so **8 = 2 x 2 x 2 = $\boxed{2^3}$**   so **10 = 2 x $\boxed{5}$**

... and then select the highest power of each prime factor that appears and form a single multiplication.

LCM of 8 and 10 is 2^3 x 5 = 40

Integers 1

What are Integers?

Integers are all the whole numbers which are greater than zero, less than zero, and zero itself. Above zero the numbers are positive although we don't write a **+** (plus) in front of them. Below zero the numbers are negative and these must have a **-** (minus) written in front of them. When you use positive and negative numbers, zero is the fixed point on the scale. All numbers relate to this point.

A number line or scale (which can be horizontal or vertical) can be a very useful aid for you to understand positive and negative numbers.

Positive and negative numbers are often used in everyday life, e.g. to show temperatures above and below freezing or financial gain and loss.

A number line

A Bank Statement showing deposits and withdrawals.

LONSDALE BUILDING SOCIETY

Date	Description	Deposit	Withdrawal	Balance
11/12/03				£226.30
12/12/03	The Toy Shop		-£49.99	£176.31
13/12/03	Gas Bill		-£21.03	£155.28
14/12/03	Cheque	£25.00		£180.28
16/12/03	La Trattoria		-£32.98	£147.30
19/12/03	Rent		-£260.00	-£112.70

Ordering Integers

Ordering means rearranging a series of positive and negative numbers in either ascending (lowest to highest) or descending (highest to lowest) order.

The simplest way is to collect all the negative and positive numbers together in two separate groups. You can then use a number line to order the numbers.

Example

Rearrange the following temperatures in ascending order:
7°C, -1°C, 2°C, 5°C, -4°C, -2°C, 3°C, -8°C

Collect the negative and positive numbers together in two separate groups and order using a number line…

… to give us the temperatures in ascending order: **-8°C, -4°C, -2°C, -1°C, 2°C, 3°C, 5°C, 7°C**

Integers 2

Addition & Subtraction of Integers

When you are adding or subtracting integers, a number line can sometimes help you. Positive numbers are counted to the right of the number line and negative numbers to the left.

Manchester
6°C

Examples...

1 At 6pm the temperature in Manchester was 6°C. By 10pm it had fallen by 8°C. The temperature at 10pm is therefore **6°C – 8°C**. On a number line this calculation can be shown by starting at 6 and then counting 8 to the **left**...

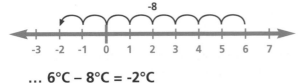

... 6°C – 8°C = -2°C

2 The temperature had fallen a further 6°C by 1am. On a number line this can be shown by starting at -2 and counting 6 places to the **left**...

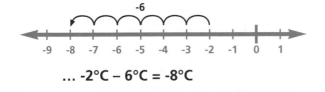

... -2°C – 6°C = -8°C

3 By 6am, the temperature had risen by 5°C. On a number line, this can be shown by starting at -8 and counting 5 places to the **right**...

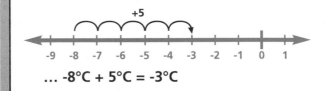

... -8°C + 5°C = -3°C

Multiplication & Division of Integers

When you are multiplying or dividing integers ignore any signs and multiply or divide the two numbers to get the number part of the answer. If the numbers have **the same** signs the answer is **positive**, whereas if the numbers have **different** signs the answer is **negative**. This table shows all the possibilities:

Multiplication of Integers	
+ X + = +	
- X - = +	
+ X - = -	
- X + = -	
Division of Integers	
+ ÷ + = +	
- ÷ - = +	
+ ÷ - = -	
- ÷ + = -	

Examples...

1 6 x -7 = -42

2 -12 x 8 = -96

3 -11 x -5 = 55

4 75 ÷ -5 = -15

5 -200 ÷ 40 = -5

6 -99 ÷ -11 = 9

Upper and Lower Bounds

Sometimes measurements are given to a certain degree of accuracy. When this happens the actual measurement (or quantity) lies somewhere in a range that has its limits a certain value below (the **lower bound**) to a certain value above (the **upper bound**).

Accuracy of given measurement	Limits of range measurement lies between
Nearest 1000	500 below to 500 above the given measurement
Nearest 100	50 below to 50 above the given measurement
Nearest 10	5 below to 5 above the given measurement
Nearest 1	0.5 below to 0.5 above the given measurement
Nearest 0.1	0.05 below to 0.05 above the given measurement

Upper and Lower Bounds for the Four Rules of Number

If **A = 50** to the nearest **10** ($45 \leqslant A < 55$) and **B = 30** to the nearest **10** ($25 \leqslant B < 35$) then…

	A + B	A - B	A x B	A ÷ B
Greatest Range	55 + 35 = 90	55 − 25 = 30	55 x 35 = 1925	55 ÷ 25 = 2.2
Smallest Range	45 + 25 = 70	45 − 35 = 10	45 x 25 = 1125	45 ÷ 35 = 1.29

Examples...

1 A sprinter runs 200m in a time of 20.2 seconds, which is measured to the nearest 0.1 of a second. What are the upper and lower bounds of the time taken?

A measurement to the nearest 0.1s means that the actual measurement lies somewhere in a range 0.05s above or below the time of 20.2s. In other words there is a possible error of ± 0.05s on the recorded time.

Lower bound = 20.2s − 0.05s = 20.15s
Upper bound = 20.2s + 0.05s = 20.25s

This can be written: **20.15s < t < 20.25s** where **t = time taken**

> The actual value can not be equal to the upper bound because if it was, the time to the nearest 0.1s would be 20.3s

2 The rectangle below has dimensions 3cm by 2cm, each length is correct to the nearest centimetre. What is the smallest and largest area possible for the rectangle?

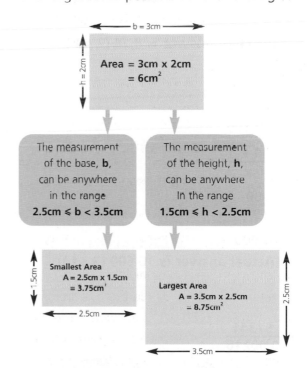

The measurement of the base, **b**, can be anywhere in the range
2.5cm ⩽ b < 3.5cm

The measurement of the height, **h**, can be anywhere in the range
1.5cm ⩽ h < 2.5cm

Smallest Area
A = 2.5cm x 1.5cm = 3.75cm²

Largest Area
A = 3.5cm x 2.5cm = 8.75cm²

Smallest Area = 3.75cm²
Largest Area = 8.75cm²

This can be written: **3.75cm² ⩽ A < 8.75cm²** where **A = area of rectangle**

Estimating and Checking

Estimating Answers

Answers to many calculations can be estimated by rounding off numbers within the calculations to 1 significant figure.

Examples...

Estimate the answers to the following calculations:

1 71 + 18 – 26

$\approx$ 70 + 20 – 30

= 60 (Actual answer is 63)

2 $\dfrac{3.6 \times 10.4}{7.7 - 3.1}$

$\approx \dfrac{4 \times 10}{8 - 3}$

$= \dfrac{40}{5}$

= 8 (Actual answer is 8.14 (2 d.p.))

3 $\dfrac{413 \times 4.87}{0.189}$

$\approx \dfrac{400 \times 5}{0.2}$

$= \dfrac{2000}{0.2}$

= 10 000
(Actual answer is 10 600 (3 s.f.))

4 $\dfrac{57341}{0.00216}$

$= \dfrac{5.7341 \times 10^4}{2.16 \times 10^{-3}}$

See page 17 for Standard Form

$\approx \dfrac{6 \times 10^4}{2 \times 10^{-3}}$

$= 3 \times 10^7$ (Actual answer is 2.65×10^7)

Checking your Answers for Accuracy

Answers to calculations can be checked for accuracy by starting with your answer and working backwards, reversing the operations.

Examples...

1 $8 \times 7 = 56 \longrightarrow \dfrac{56}{8} = 7 \checkmark$ or $\dfrac{56}{7} = 8 \checkmark$

2

The table below shows the amount of money collected by a small local charity from October to December, using collecting tins at three different locations. Check it for accuracy.

MONTH	Location 1	Location 2	Location 3	TOTAL
Oct	28.08	16.73	18.23	63.04
Nov	32.96	21.01	18.16	72.13
Dec	26.12	19.57	16.03	61.72
TOTAL	87.16	57.31	52.42	**196.89**

For this example, there are several ways of checking it for accuracy. You could check the totals for each month (rows) or the totals for each location (columns). Finally, check the overall total.

- **63.04 + 72.13 + 61.72 = 196.89**
  196.89 – 61.72 – 72.13 = 63.04 $\checkmark$

- **87.16 + 57.13 + 52.42 = 196.89**
  196.89 – 52.42 – 57.31 = 87.16 $\checkmark$

Bidmas

The simplest possible calculation involves only one operation, e.g. an addition or multiplication.

However, when a calculation involves more than one operation, you must carry them out in the order shown here.

B I D M A S

| BRACKETS | INDICES (OR POWERS) | DIVISIONS AND MULTIPLICATIONS - THESE CAN BE DONE IN ANY ORDER | ADDITIONS AND SUBTRACTIONS - THESE CAN BE DONE IN ANY ORDER |

Examples...

1 $8 + 3 \times 4$

Do the multiplication first

$= 8 + 12$

Then the addition

$= 20$

2 $\dfrac{(14 + 6)}{-4}$

Do the addition in the bracket first

$= \dfrac{20}{-4}$

Then the division

$= -5 \; (+ \div - = -)$

3 $4^2 - 2 \times 5$

Work out the square first

$= 16 - 2 \times 5$

Then the multiplication

$= 16 - 10$

Then the subtraction

$= 6$

Work through these examples again, only this time do the operations in the wrong order. You should get different answers which are wrong!

Use of Brackets

The insertion of a bracket (or brackets) into a calculation can change the final answer.

Do remember that if you are asked to insert a bracket into a calculation that involves more than one operation there is always more than one result. Make sure that you try out all the different results.

Example...

Here is a calculation without brackets.

$8 + 3 \times 4 - 2 = 8 + 12 - 2 = 20 - 2 = 18$

(or $8 + 10$)

If we now insert one bracket into the calculation the possible answers become...

$(8 + 3) \times 4 - 2 = 11 \times 4 - 2 = 44 - 2 = 42$

$8 + 3 \times (4 - 2) = 8 + 3 \times 2 = 8 + 6 = 14$

Powers 1

Understanding Powers

Powers or indices show that a number is to be multiplied by itself a certain number of times.

$4 \times 4 \qquad = 4^2$ (4 squared)

$4 \times 4 \times 4 \qquad = 4^3$ (4 cubed)

$4 \times 4 \times 4 \times 4 \qquad = 4^4$ (4 to the power 4)

$4 \times 4 \times 4 \times 4 \times 4 = 4^5$ (4 to the power 5)

... and so on

$$4^2 \leftarrow \text{the power or index}$$

Powers of Negative Numbers

Care is needed when working out powers of negative numbers. If you are in any doubt refer back to page 10 for the multiplication of integers.

Examples...

1 $(-4)^2 = -4 \times -4 = 16$

(since a 'minus' times a 'minus' is equal to a 'plus')

2 $(-4)^3 = -4 \times -4 \times -4 = -64$

(the first two 'minuses' give a 'plus'. This 'plus' times a 'minus' then gives a 'minus').

Square Numbers

Numbers obtained by squaring a number are called square numbers, for example...

4 squared $= 4^2 = 4 \times 4 = 16$

9 squared $= 9^2 = 9 \times 9 = 81$

The first four square numbers are...

1	**4**	**9**	**16**
$(1^2 = 1 \times 1)$	$(2^2 = 2 \times 2)$	$(3^2 = 3 \times 3)$	$(4^2 = 4 \times 4)$

You are expected to be able to recall integer squares from 2^2 to 15^2.

$2^2=$	$3^2=$	$4^2=$	$5^2=$	$6^2=$	$7^2=$	$8^2=$
4	9	16	25	36	49	64

$9^2=$	$10^2=$	$11^2=$	$12^2=$	$13^2=$	$14^2=$	$15^2=$
81	100	121	144	169	196	225

Cube Numbers

Numbers obtained by cubing a number are called cube numbers, for example...

4 cubed $= 4^3 = 4 \times 4 \times 4 = 64$

9 cubed $= 9^3 = 9 \times 9 \times 9 = 729$

The first four cube numbers are...

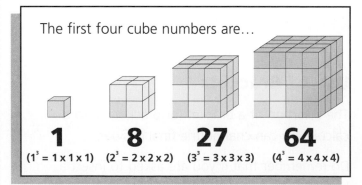

1	**8**	**27**	**64**
$(1^3 = 1 \times 1 \times 1)$	$(2^3 = 2 \times 2 \times 2)$	$(3^3 = 3 \times 3 \times 3)$	$(4^3 = 4 \times 4 \times 4)$

This time you are expected to be able to recall the cubes of the following numbers:

$2^3=$	$3^3=$	$4^3=$	$5^3=$	$10^3=$
8	27	64	125	1000

Rules of Indices

1 Powers are added when we multiply powers of the same number.
$$3^5 \times 3^2 = 3^{5+2} = 3^7, \; 10^3 \times 10 = 10^{3+1} = 10^4$$

2 Powers are subtracted when we divide powers of the same number.
$$3^5 \div 3^2 = 3^{5-2} = 3^3, \; 10^3 \div 10 = 10^{3-1} = 10^2$$

3 Powers are multiplied when a power of a number is raised to another power.
$$(3^5)^2 = 3^{5\times2} = 3^{10}, \; (10^3)^1 = 10^{3\times1} = 10^3$$

Also…

4 A number to a power of 1 is the number itself.
$3^1 = 3$ and vice versa $3 = 3^1$
$10^1 = 10$ and vice versa $10 = 10^1$

5 A number to a power of 0 is always equal to 1.
$3^0 = 1$, $10^0 = 1$

6 A number to a negative power is one over that number to the power.
$3^{-2} = \dfrac{1}{3^2}$ and vice versa $\dfrac{1}{3^2} = 3^{-2}$
$10^{-1} = \dfrac{1}{10^1}$ and vice versa $\dfrac{1}{10^1} = 10^{-1}$

Fractional Powers

A fractional power simply means a square root, cube root, etc. (see page 16).

1 $4^{\frac{1}{2}}$ means $\sqrt{4} = \pm2$…
… a power of $\frac{1}{2}$ means square root.

2 $27^{\frac{1}{3}}$ means $\sqrt[3]{27} = 3$…
… a power of $\frac{1}{3}$ means cube root.

3 $10\,000^{\frac{1}{4}}$ means $\sqrt[4]{10\,000} = \pm10$…
… a power of $\frac{1}{4}$ means fourth root and so on.

Also…
The inverse (opposite) operation of raising a positive number to power 'n' is raising the result of this operation to power '$\frac{1}{n}$', for example…

$4^2 = 16$ and $16^{\frac{1}{2}} = 4$, $5^3 = 125$ and $125^{\frac{1}{3}} = 5$

Examples…

1 Evaluate $\dfrac{3^4 \times 3^7}{3^8}$

$= \dfrac{3^{4+7}}{3^8}$ — Do the multiplication first.
— Then the division.

$= 3^{11-8}$

$= 3^3$

$= 27$

2 Evaluate $\dfrac{(2^4)^2}{2^{-5}}$

$= \dfrac{2^{4\times2}}{2^{-5}}$ — Do the power raised to another power first.
— Then the division. The same rules still apply for negative powers as for positive powers.

$= 2^{8--5}$

$= 2^{8+5}$

$= 2^{13}$

3 Evaluate $\dfrac{4^7 \times 4^{-2}}{4^5}$

$= \dfrac{4^{7+-2}}{4^5}$ — Do the multiplication first.
— Then the division.

$= 4^{5-5}$

$= 4^0$

$= 1$

(since any number to the power of 0 is equal to 1)

1 Calculate $8^{\frac{2}{3}}$

In this case we need to split the fractional power up into two powers,
e.g. $\frac{2}{3} = \frac{1}{3} \times 2$. We can now do the calculation in two stages. Therefore…

$$8^{\frac{2}{3}} = 8^{\frac{1}{3} \times 2} = 2^2 = 4$$
stage 1 / stage 2

Cube root — Square
of 8 = 2 — of 2 = 4

— Always do the root first as this decreases the size of your number, and then the power.

Calculate $32^{\frac{3}{5}}$

2 $32^{\frac{3}{5}} = 32^{\frac{1}{5} \times 3} = 2^3 = 8$
stage 1 / stage 2

Fifth root — Cube of
of 32 = 2 — 2 = 8

Square Roots of Positive Numbers

You know that…

$$4^2 = 4 \times 4 = 16$$

and also

$(-4)^2 = -4 \times -4 = 16$ (see page 14)

Another way of describing the above is to say that **+4** or **-4 is the square root of 16**. Therefore…

$\sqrt{16}$ or $16^{\frac{1}{2}} = \pm 4$ (i.e. +4 or -4)

This means that whenever you find the square root of a positive number there will always be a positive square root and a negative square root, for example…

$\sqrt{25}$ or $25^{\frac{1}{2}} = \pm 5$ (i.e. +5 or -5)

$\sqrt{36}$ or $36^{\frac{1}{2}} = \pm 6$ (i.e. +6 or -6)

You will be expected to be able to recall the square roots of all integer squares from 2^2 **(4)** to 15^2 **(225)**.

$\sqrt{4}=$	$\sqrt{9}=$	$\sqrt{16}=$	$\sqrt{25}=$	$\sqrt{36}=$	$\sqrt{49}=$	$\sqrt{64}=$
±2	±3	±4	±5	±6	±7	±8

$\sqrt{81}=$	$\sqrt{100}=$	$\sqrt{121}=$	$\sqrt{144}=$	$\sqrt{169}=$	$\sqrt{196}=$	$\sqrt{225}=$
±9	±10	±11	±12	±13	±14	±15

Cube Roots of Positive Numbers

Finding the cube root of a positive number results in only one root … a positive root.

Since $4^3 = 4 \times 4 \times 4 = 64$,

then **4 is the cube root of 64**. Therefore…

$\sqrt[3]{64}$ or $64^{\frac{1}{3}} = 4$

Similarly…

$\sqrt[3]{125}$ or $125^{\frac{1}{3}} = 5$ **(since 5 x 5 x 5 = 125)**

$\sqrt[3]{1000}$ or $1000^{\frac{1}{3}} = 10$ **(since 10 x 10 x 10 = 1000)**

Surds

Some numbers (square numbers) have an exact square root, e.g. $\sqrt{4} = 2$. Other numbers do not, e.g. $\sqrt{5} = 2.236067\ldots$ Since $\sqrt{5}$ cannot be written as an exact number then it is more accurate to leave it as $\sqrt{5}$. When it is written in this form, $\sqrt{5}$ is called a surd. Many surds can be simplified…

1 $\sqrt{18} = \sqrt{9 \times 2} = \sqrt{9} \times \sqrt{2} = 3 \times \sqrt{2} = 3\sqrt{2}$

2 $\sqrt{48} = \sqrt{16 \times 3} = \sqrt{16} \times \sqrt{3} = 4 \times \sqrt{3} = 4\sqrt{3}$

3 $\dfrac{\sqrt{10}}{\sqrt{2}} = \sqrt{\dfrac{10}{2}} = \sqrt{5}$

4 $(4 - \sqrt{3})^2 = (4 - \sqrt{3})(4 - \sqrt{3})$
$= 4 \times 4 + 4 \times -\sqrt{3} - \sqrt{3} \times 4 - \sqrt{3} \times -\sqrt{3}$
$= 16 - 4\sqrt{3} - 4\sqrt{3} + 3$
$= 19 - 8\sqrt{3}$

Decimal numbers that are neither terminating nor recurring (see page 22) are known as irrational numbers. All other numbers are rational. For example, the square root of 5 is 2.236067… a decimal number that does not terminate or recur. This is an example of an irrational number.

The following example shows you how to rationalise a denominator.

$$\frac{2}{\sqrt{5}} = \frac{2 \times \sqrt{5}}{\sqrt{5} \times \sqrt{5}} = \frac{2\sqrt{5}}{5}$$

Multiply the numerator (top number) and denominator (bottom number) by the denominator itself, i.e. $\sqrt{5}$. The denominator is now a rational number.

Standard Index Form

What is Standard Index Form?

If you are dealing with very large or very small numbers, it is not always practical to write them out in full each time. Standard Index Form, or Standard Form, is an alternative way of writing very large or very small numbers in shorthand.

To write a number in standard index form, you move its decimal point a certain number of places so that the number is written as $\mathbf{a \times 10^n}$ … where **a** must be a number that is equal to or greater than **1** and less than **10**, and **n** is a positive or negative integer.

For **large** numbers greater than **1**, **n** is a **positive integer** and is equal to the number of places the decimal point has moved.

Examples...

1 $4372 = 4.372 \times 10^3$
 4.372 — The decimal point has moved 3 places

2 $691000 = 6.91 \times 10^5$
 6.91000 — The decimal point has moved 5 places

Numbers written in standard index form can also be converted to normal numbers...

3 $3.71 \times 10^4 = 37100$
 37100.

For **small** positive numbers less than **1**, **n** is a **negative integer** (**min**us for **min**ute numbers) and is equal to the number of places the decimal point has moved.

Examples...

1 $0.0356 = 3.56 \times 10^{-2}$
 003.56 — The decimal point has moved 2 places

2 $0.00013 = 1.3 \times 10^{-4}$
 00001.3 — The decimal point has moved 4 places

Numbers written in standard index form can also be written as normal numbers...

3 $4.5712 \times 10^{-5} = 0.000045712$
 .000045712

Calculations with Standard Index Form

To ADD or SUBTRACT numbers written in standard index form, convert to 'normal numbers', do the calculation and then, if asked, change your answer back into standard index form.

To MULTIPLY or DIVIDE numbers written in standard index form, carry out separate calculations for the numbers and powers and then rearrange your answer back into standard index form.

Examples...

1 $\quad 4.2 \times 10^4 + 8.6 \times 10^3$
$= 42000 + 8600$
$= 50600$
$= 5.06 \times 10^4$

2 $\quad 9.37 \times 10^{-2} - 1.6 \times 10^{-3}$
$= 0.0937 - 0.0016$
$= 0.0921$
$= 9.21 \times 10^{-2}$

Examples...

1 $\quad 3.2 \times 10^{-3} \times 4.5 \times 10^{-4}$
$= (3.2 \times 4.5) \times (10^{-3} \times 10^{-4})$
$= 14.4 \times 10^{-7}$
$= 1.44 \times 10^1 \times 10^{-7} = 1.44 \times 10^{-6}$

2 $\quad 5.6 \times 10^5 \div 8 \times 10^1$
$= (5.6 \div 8) \times (10^5 \div 10^1)$
$= 0.7 \times 10^4$
$= 7 \times 10^{-1} \times 10^4 = 7 \times 10^3$

Fractions 1

Understanding Simple Fractions

The diagram shows a whole pizza that has been cut into four equal parts (or slices). Each slice can be described as a fraction of the whole pizza and its value is $\frac{1}{4}$.

The pizza however could have been cut into any number of slices, where each slice is a different fraction of the whole pizza (see the diagrams below).

The top number of a fraction is called the **numerator** while the bottom number is called the **denominator**.

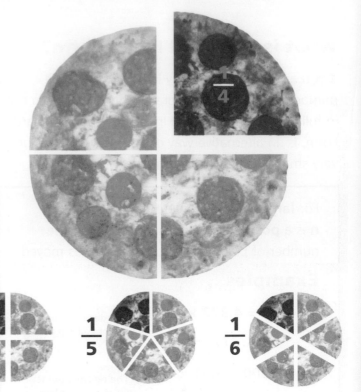

$\frac{1}{2}$ $\frac{1}{3}$ $\frac{1}{4}$ $\frac{1}{5}$ $\frac{1}{6}$

Equivalent Fractions

These are fractions that are equal. You can build chains of equivalent fractions by multiplying the numerator and denominator in the fraction by the same number.

$$\frac{1}{4} = \frac{2}{8} = \frac{4}{16} = \frac{8}{32}$$

x 2   x 2   x 2

Cancelling Fractions

You can also divide the numerator and denominator in a fraction by the same number. Chains of equivalent fractions that are cancelled down using division always come to an end when the fraction is expressed in its simplest form, i.e. lowest terms.

$$\frac{18}{27} = \frac{6}{9} = \frac{2}{3} \quad \text{or} \quad \frac{18}{27} = \frac{2}{3}$$

÷ 3   ÷ 3   ÷ 9

Ordering Fractions

Example...

Arrange $\frac{9}{10}$, $\frac{4}{5}$, $\frac{7}{8}$ in ascending order.

Firstly write each fraction with a common denominator. The lowest common multiple of 10, 5 and 8 is 40. Therefore…

$$\frac{9}{10} = \frac{36}{40} \quad \frac{4}{5} = \frac{32}{40} \quad \frac{7}{8} = \frac{35}{40}$$

x 4   x 8   x 5

If we now compare the numerators we can arrange the fractions in ascending (lowest to highest) order. In ascending order they are: $\frac{4}{5}$, $\frac{7}{8}$, $\frac{9}{10}$

Also when you have fractions with common denominators it is possible to find other fractions that lie between them.

For example, from above $\frac{4}{5} = \frac{32}{40}$ and $\frac{7}{8} = \frac{35}{40}$. We can write two other fractions with denominator 40 which are greater than $\frac{4}{5}$ but less than $\frac{7}{8}$. They would be $\frac{33}{40}$ and $\frac{34}{40}$, which cancels down to $\frac{17}{20}$.

Addition of Fractions

Two fractions can be added very easily providing they have common denominators.

$$\frac{2}{5} = \frac{8}{20} \quad \frac{2}{5} + \frac{3}{4} \quad \frac{3}{4} = \frac{15}{20}$$
$$= \frac{8}{20} + \frac{15}{20}$$
$$= \frac{23}{20}$$
$$= 1\frac{3}{20}$$

With mixed numbers, add the whole numbers and fractions separately and then combine.

$$2\frac{2}{3} + 3\frac{1}{7}$$
$$= (2 + 3) + \left(\frac{2}{3} + \frac{1}{7}\right)$$
$$= 5 + \left(\frac{14}{21} + \frac{3}{21}\right)$$
$$= 5 + \frac{17}{21} = 5\frac{17}{21}$$

Subtraction of Fractions

As for addition, two fractions can be subtracted very easily providing they have common denominators.

$$\frac{7}{8} = \frac{21}{24} \quad \frac{7}{8} - \frac{2}{3} \quad \frac{2}{3} = \frac{16}{24}$$
$$= \frac{21}{24} - \frac{16}{24}$$
$$= \frac{5}{24}$$

With mixed numbers, subtract the whole numbers and fractions separately and then combine.

$$4\frac{1}{2} - 1\frac{4}{5}$$
$$= (4 - 1) + \left(\frac{1}{2} - \frac{4}{5}\right)$$
$$= 3 + \left(\frac{5}{10} - \frac{8}{10}\right)$$
$$= 3 + \left(-\frac{3}{10}\right)$$
$$= 3 - \frac{3}{10}$$
$$= \left(2 + \frac{10}{10}\right) - \frac{3}{10} = 2\frac{7}{10}$$

Multiplication and Division of Fractions

To multiply two fractions, multiply together the two numerators and the two denominators.

$$\frac{2}{3} \times \frac{4}{5} = \frac{2 \times 4}{3 \times 5} = \frac{8}{15}$$

Division of two fractions is the same as multiplication, except that you turn the second fraction (that is doing the dividing) upside down and change the division sign to a multiplication sign.

$$\frac{2}{3} \div \frac{4}{5} = \frac{2}{3} \times \frac{5}{4} = \frac{2 \times 5}{3 \times 4} = \frac{10}{12} = \frac{5}{6}$$

To multiply or divide mixed numbers you have to convert them to improper fractions first.

$$2\frac{1}{2} \times 1\frac{1}{6} = \frac{5}{2} \times \frac{7}{6} = \frac{5 \times 7}{2 \times 6} = \frac{35}{12} = 2\frac{11}{12}$$

$$1\frac{1}{3} \div 3\frac{1}{2} = \frac{4}{3} \div \frac{7}{2} = \frac{4}{3} \times \frac{2}{7} = \frac{4 \times 2}{3 \times 7} = \frac{8}{21}$$

To multiply or divide a fraction by an integer, convert the integer to an improper fraction. For example, **4** as an improper fraction is $\frac{4}{1}$. You can then do the multiplication or division as normal.

$$\frac{4}{7} \times 3 = \frac{4}{7} \times \frac{3}{1} = \frac{4 \times 3}{7 \times 1} = \frac{12}{7} = 1\frac{5}{7}$$

$$\frac{2}{3} \div 10 = \frac{2}{3} \div \frac{10}{1} = \frac{2}{3} \times \frac{1}{10} = \frac{2 \times 1}{3 \times 10} = \frac{2}{30} = \frac{1}{15}$$

Calculations Involving Fractions

Calculating a Fraction of a Quantity

To find a fraction of any quantity, you have to multiply the fraction by the quantity. In other words, 'of' means 'times' or 'multiply' (x).

£9,000

Examples...

① Calculate $\frac{4}{5}$ of 60kg.

'of' means 'x'

$$\frac{4}{5} \text{ of 60kg} = \frac{4}{5} \times 60$$

$$= \frac{4}{5} \times \frac{60}{1}$$

$$= \frac{240}{5}$$

$$= 48\text{kg}$$

② A brand new car costing £9 000 will lose $\frac{1}{5}$ of its value in the first year. What is the value of the car after the first year?

Before we can calculate its value we need to calculate the loss.

'of' means 'x'

Loss = $\frac{1}{5}$ of £9000 = $\frac{1}{5}$ x 9000 = $\frac{1}{5}$ x $\frac{9000}{1}$ = $\frac{9000}{5}$ = £1800

Value of car after the first year = £9000 – £1800 = £7200.

OR 1 – $\frac{1}{5}$ = $\frac{4}{5}$ of value after first year = £9000 x $\frac{4}{5}$ = $\frac{36\,000}{5}$ = £7200.

Expressing One Quantity as a Fraction of Another Quantity

Firstly, you must make sure that both quantities are in the same units. Then, to express the relationship as a fraction, the first quantity becomes the numerator (top number) and the second quantity becomes the denominator (bottom number). If need be, write the fraction in its lowest terms.

Examples...

① Write 30 out of 120 as a fraction.

30 out of 120 = $\frac{30}{120}$ = $\frac{1}{4}$

② Express 36 seconds as a fraction of 2 minutes. Both quantities must be in the same units, so change 2 minutes into seconds.

2 minutes = 2 x 60s = 120s, so...

36s as a fraction of 120s = $\frac{36}{120}$ = $\frac{3}{10}$

Understanding Simple Decimals

When you use money and metric measures, such as millimetres, litres or grams, you often use decimals. Decimals are an easy kind of fraction - easy because the ten 'times' table is the only one ever needed! If you think of a pizza, with decimals the whole pizza is only ever divided into ten, a hundred or a thousand pieces.

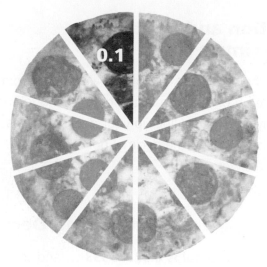

where each slice = $\frac{1}{10}$ or **0.1**

0.1  **0.2**  **0.3**  **0.4**  **0.5**  **etc.**

Place Value in Decimal Numbers

We have already seen that all the digits in a whole number have a place value. In a decimal number, all digits to the right of the decimal point also have a place value.

Decimal Number	Place Value of Digits					
	10 (Tens)	1 (Units)	DECIMAL POINT	$\frac{1}{10}$ (Tenths)	$\frac{1}{100}$ (Hundredths)	$\frac{1}{1000}$ (Thousandths)
0.06		0		0	6	
0.507		0		5	0	7
1.39		1		3	9	
74.258	7	4		2	5	8

The position of the decimal point is just as important as the digits themselves. Misplacing the decimal point makes a huge difference because all the place values change. For example, in a long jump contest a pupil records a jump of 4.37m. Imagine the excitement if the distance was recorded as 43.7m!

Ordering Decimals

This means rearranging a series of decimals in either ascending (lowest to highest) or descending (highest to lowest) order. A useful method is to line up all the decimal points of the numbers in a vertical line. Then you can start with the first column on the left, and work along each column of numbers (left to right) to decide which is the biggest number.

Example...

Rearrange the following in ascending order: 0.54, 5.4, 0.45, 4.5

Line up the decimal points.

Decimals 2

Addition and Subtraction of Decimals

As for whole numbers the place values of the digits must line up one on top of the other, although an easy way is to simply line up your decimal points. Remember to bring the decimal point down to your answer.

Examples...

```
    343:62
  +  17:59
     ¹ ¹  ¹
    361:21
```

```
    5 1 0 1
   6̸3:1̸84
  -  5:091
    58:093
```

Recurring and Terminating Decimals

$\frac{1}{3}$ **as a decimal is 0.33333333 ...** and so on

$\frac{3}{11}$ **as a decimal is 0.27272727 ...** and so on

Both of these are examples of recurring decimals because one or more of the digits repeats itself continuously. To make it simpler we place a dot (•) over the digit or digits that repeat continuously.

$\frac{1}{3} = 0.33333333 ... = 0.\dot{3}$

$\frac{3}{11} = 0.27272727 ... = 0.\dot{2}\dot{7}$

Decimals that do not recur are called terminating decimals. Fractions, written in their **simplest form**, will convert into terminating decimals **if** their denominators have prime factors of either **2** or **5** or **both**.

$\frac{7}{10} = 0.7$ (since $10 = 2 \times 5$)

$\frac{5}{8} = 0.625$ (since $8 = 2 \times 2 \times 2$)

$\frac{19}{50} = 0.38$ (since $50 = 2 \times 5 \times 5$)

Converting Recurring Decimals into Fractions

Examples...

1 **0.3333...** the recurring pattern occurs after the FIRST NUMBER

> Let our recurring decimal be represented by the letter N

> If the recurring pattern occurs after the first number, multiply by 10.

$N = 0.3333...$

$10N = 0.3333... \times 10$

$10N = 3.3333...$

$10N - N = 3.3333... - 0.3333...$

$9N = 3$

$N = \frac{3}{9} = \frac{1}{3}$

$0.3333... = \frac{1}{3}$

2 **0.2727...** the recurring pattern occurs after the SECOND NUMBER

> Let our recurring decimal be represented by the letter N

> If the recurring pattern occurs after the second number, multiply by 100 and so on...

$N = 0.2727...$

$100N = 0.2727... \times 100$

$100N = 27.2727...$

$100N - N = 27.2727... - 0.2727...$

$99N = 27$

$N = \frac{27}{99} = \frac{3}{11}$

$0.2727... = \frac{3}{11}$

Multiplication With Decimals

Multiplication of Decimal Numbers by Powers of 10

To multiply a decimal number by a power of 10, e.g. 10 (10^1), 100 (10^2), 1000 (10^3), all you have to do is move all the digits a certain number of place values to the left. When you do this your number becomes bigger. For example…

1 $9.32 \times 10 = 93.2$

Digits move one place value to the left and the number becomes 10 times bigger.

2 $0.047 \times 100 = 4.7$

Digits move two place values to the left and the number becomes 100 times bigger.

3 $13.27 \times 1\,000 = 13\,270$

Digits move three place values to the left and the number becomes 1000 times bigger.

Multiplication of Positive Numbers by Decimal Numbers between 0 and 1

If you multiply any positive number by a decimal number between 0 and 1 the answer is always smaller than the positive number you started with, for example…

1 $10 \times 0.5 = 5$

2 $5 \times 0.3 = 1.5$

3 $986 \times 0.01 = 9.86$

Multiplication of Decimal Numbers by Whole and Decimal Numbers

Ignore the decimal points and multiply as you would whole numbers. You put the decimal point in at the end. In your answer, the number of digits after the decimal point should be the same as the total number of digits after the decimal points in the numbers being multiplied. For example…

1 2.73×18

Multiply as you would whole numbers.

```
      273
 x     18
     2730
   2184
   4914
```

Then, count the digits after the decimal point in the numbers being multiplied and transfer to answer.

$2.73 \times 18 = 49.14$

Therefore $2.73 \times 18 = 49.14$

2 17.5×0.61

Multiply as you would whole numbers.

```
      175
 x     61
    10500
      175
    10675
```

Then, count the digits after the decimal point in the numbers being multiplied and transfer to answer.

$17.5 \times 0.61 = 10.675$

Therefore $17.5 \times 0.61 = 10.675$

Division With Decimals

Division of Decimal Numbers by Powers of 10

The process is the reverse of multiplying decimal numbers by powers of 10. With division all the digits move a certain number of place values to the right. When you do this your number becomes smaller. For example…

1 **46.3 ÷ 10 = 4.63**

Digits move one place value to the right and the number becomes 10 times smaller.

2 **3.615 ÷ 100 = 0.03615**

Digits move two place values to the right and the number becomes 100 times smaller.

3 **473.2 ÷ 1000 = 0.4732**

Digits move three place values to the right and the number becomes 1000 times smaller.

Division of Positive Numbers by Decimal Numbers between 0 and 1

If you divide any positive number by a decimal number between 0 and 1 the answer is always bigger than the positive number you started with, for example…

1 **10 ÷ 0.5 = 20**

2 **5 ÷ 0.4 = 12.5**

3 **471 ÷ 0.01 = 47 100**

Division of Decimal Numbers by Whole Numbers and Decimal Numbers

Division of a decimal number by a whole number is the same as the division of whole numbers (see page 6). The only exception is that you must remember to take the decimal point up to your answer.

13.2 ÷ 6

Divide as you would whole numbers.

```
      2 . 2
  6 ) 1 3 ⸍ 2
      1 2   ↓
      ‾‾‾‾
        1 2
        1 2
        ‾‾‾
          0
```

Remember to take the decimal point up to the answer.

Therefore **13.2 ÷ 6 = 2.2**

Division of a decimal number by another decimal number is slightly more tricky. Before you start, multiply both numbers by 10, 100 etc. until the number doing the dividing is a whole number. The process from now on is the same as the example above.

4.368 ÷ 0.56

Multiply both numbers by 100 to make the divisor a whole number. **4.368 x 100 = 436.8, 0.56 x 100 = 56**

Divide as you would whole numbers.

```
         7 . 8
  5 6 ) 4 3 6 ⸍ 8
       3 9 2   ↓
       ‾‾‾‾‾
         4 4   8
         4 4   8
         ‾‾‾‾‾
             0
```

Remember to take the decimal point up to the answer.

Therefore **4.368 ÷ 0.56 = 7.8**

Percentages 1

Understanding Simple Percentages

Percentages are used in everyday life, from pay rises to price reductions. This helps us to make easy comparisons. When you understand the basics of percentages they are even easier than decimals or fractions. They focus on the whole being equal to one hundred, i.e. the whole is one hundred per cent. So, if someone gives you 50% of a whole pizza this tells you the 'number of parts per 100' you have, e.g. 50% is 50 parts per 100 or $\frac{50}{100}$

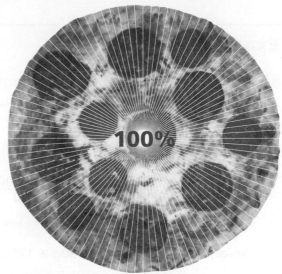

100%

10% 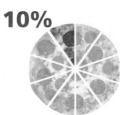 **20%** 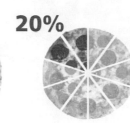 **30%** 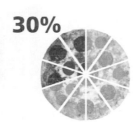 **40%** 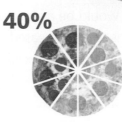 **50%**  **etc.**

Calculating a Percentage of a Quantity

Examples...

❶ Calculate 40% of 50cm.

40% means 40 parts per 100 or $\frac{40}{100}$ and 'of' means 'times' or 'multiply'(x).

40% of 50cm = $\frac{40}{100}$ x 50 = $\frac{40 \times 50}{100}$

= 20cm

❷ Calculate 15% of £4.80
Give your answer in pence.

£4.80 = 4.80 x 100p = 480p

15% means 15 parts out of 100 or $\frac{15}{100}$ and 'of' means 'times' (x).

15% of 480p = $\frac{15}{100}$ x 480

= $\frac{15 \times 480}{100}$

= 72p

Expressing One Quantity as a Percentage of Another Quantity

Make sure that both quantities are in the same units. Firstly, express one quantity as a fraction of the other, where the first quantity becomes the numerator (top number) and the second quantity becomes the denominator (bottom number). Then multiply the fraction by 100%.

Examples...

❶ Write 18 out of 30 as a percentage.

18 out of 30 = $\frac{18}{30}$ x 100%

= $\frac{18 \times 100}{30}$ = 60%

❷ Express 30cm as a percentage of 3m.
Both quantities must be in the same units, so change 3m into centimetres.

3m = 3 x 100cm = 300cm, so...

30cm as a % of 300cm = $\frac{30}{300}$ x 100%

= $\frac{30 \times 100}{300}$ = 10%

Percentages 2

Examples...

1 A standard box of breakfast cereal weighs 500g. Special boxes contain an extra 25%. Calculate the weight of a special box of cereal.

> Firstly calculate the increase in weight...

$$25\% \text{ of } 500g = \frac{25}{100} \times 500g = \frac{25 \times 500}{100} = 125g$$

> ... then add it to the weight of a standard box of cereal.

Weight of special box = 500g + 125g = 625g

An alternative method would be as follows: A standard box of cereal is to be increased by 25%. If a standard box is 100%, a special box will be 100% + 25% = 125%.

$$125\% = \frac{125}{100} = 1.25$$

> In other words, the weight of a special box is 1.25 times the weight of a standard box.

Weight of special box = 1.25 x 500g = 625g.

2 On 1st January John's weight is 80kg. In the first six months of the year his weight increases by 10%, followed by a 10% decrease in the second six months of the year. What is his weight at the end of the year?

After the first six months, his weight is **100% + 10% = 110%** of his original weight.

$$110\% = \frac{110}{100} = 1.1$$

> i.e. his weight is now 1.1 times its original amount.

After the second six months, his weight is **100% – 10% = 90%** of his weight at the end of the first six months.

$$90\% = \frac{90}{100} = 0.9$$

> i.e. his weight becomes 0.9 times the previous amount, which is actually a decrease.

Weight at end of year = 1.1 x 0.9 x 80kg = 79.2kg

3 A bouncing ball is dropped from a height of 5m. It reaches 80% of its previous height after every bounce. How high does it reach after the third bounce?

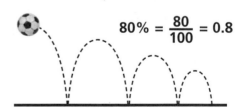

$$80\% = \frac{80}{100} = 0.8$$

> So, the height the ball reaches is 0.8 times the previous height reached.

After 3 bounces, height reached = 0.8 x 0.8 x 0.8 x 5m = 2.56m

Alternatively, this is an example of a repeated multiplier. In other words $0.8 \times 0.8 \times 0.8 = 0.8^3$ (0.8 to the power 3) to give us...

After 3 bounces, height reached = 0.8^3 x 5m = 0.512 x 5m = 2.56m

Calculations Involving Comparisons

These are calculations where the original amount has been increased or decreased by a certain percentage to a transformed amount. You must remember that the original amount is always equal to 100% and that the transformed amount will always be a percentage that is above or below 100%.

Examples...

1 A shop offers 10% off all its clothes. A shirt has a sale price of £27. What was the original price of the shirt?

Original price of shirt = 100%
Sale price of shirt = 100% - 10%

There has been a decrease of 10% → **= 90%**

$$\therefore \quad \begin{matrix} \div 90 \\ \times 100 \end{matrix} \left\{ \begin{matrix} 90\% = £27 \\ 1\% = £0.30 \\ 100\% = £30 \end{matrix} \right\} \begin{matrix} \div 90 \\ \times 100 \end{matrix}$$

So the original price of the shirt was £30

2 In 2006 a company allocated £15 000 of its budget to be spent on advertising. This was a 20% increase on the money spent on advertising in 2005. How much money did the company spend on advertising in 2005?

Original amount in this question is the money spent on advertising in 2005.

Money spent on advertising in 2005 = 100%
Money spent on advertising in 2006 = 100% + 20%

There has been an increase of 20% → **= 120%**

$$\therefore \quad \begin{matrix} \div 120 \\ \times 100 \end{matrix} \left\{ \begin{matrix} 120\% = £15\,000 \\ 1\% = £125 \\ 100\% = £12\,500 \end{matrix} \right\} \begin{matrix} \div 120 \\ \times 100 \end{matrix}$$

So the advertising budget for 2005 was £12 500

2005

2006

Converting Between Systems

Below is a summary of how to convert between fractions, decimals and percentages. You may start at any of the three points. Trace around the charts and see how you can move from one system to another by following simple rules.

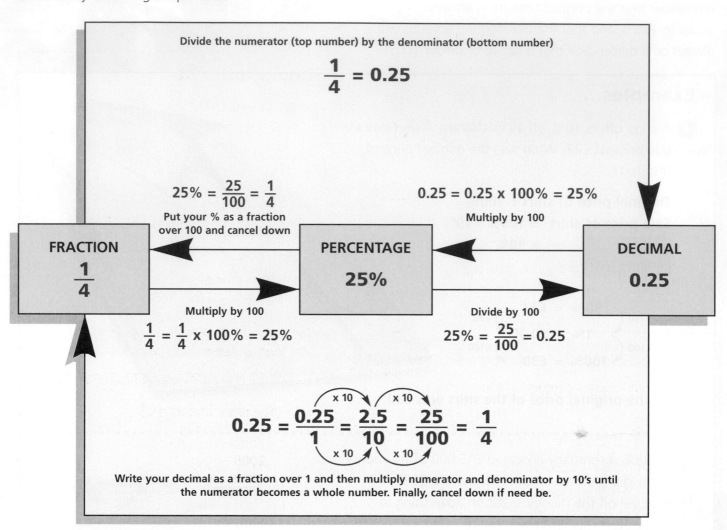

Divide the numerator (top number) by the denominator (bottom number)

$$\frac{1}{4} = 0.25$$

$25\% = \frac{25}{100} = \frac{1}{4}$
Put your % as a fraction over 100 and cancel down

$0.25 = 0.25 \times 100\% = 25\%$
Multiply by 100

| **FRACTION** $\frac{1}{4}$ | | **PERCENTAGE** 25% | | **DECIMAL** 0.25 |

Multiply by 100
$\frac{1}{4} = \frac{1}{4} \times 100\% = 25\%$

Divide by 100
$25\% = \frac{25}{100} = 0.25$

$$0.25 = \frac{0.25}{1} = \frac{2.5}{10} = \frac{25}{100} = \frac{1}{4}$$

Write your decimal as a fraction over 1 and then multiply numerator and denominator by 10's until the numerator becomes a whole number. Finally, cancel down if need be.

Other Examples...

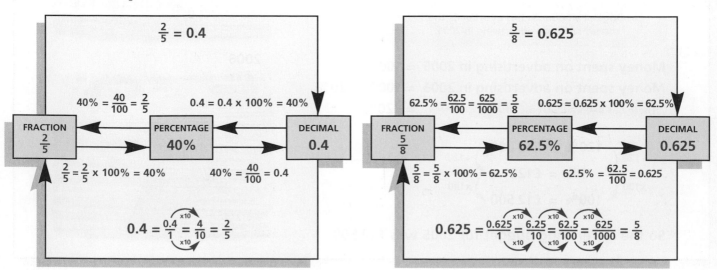

$$\frac{2}{5} = 0.4$$

$40\% = \frac{40}{100} = \frac{2}{5}$ $0.4 = 0.4 \times 100\% = 40\%$

| **FRACTION** $\frac{2}{5}$ | | **PERCENTAGE** 40% | | **DECIMAL** 0.4 |

$\frac{2}{5} = \frac{2}{5} \times 100\% = 40\%$ $40\% = \frac{40}{100} = 0.4$

$$0.4 = \frac{0.4}{1} = \frac{4}{10} = \frac{2}{5}$$

$$\frac{5}{8} = 0.625$$

$62.5\% = \frac{62.5}{100} = \frac{625}{1000} = \frac{5}{8}$ $0.625 = 0.625 \times 100\% = 62.5\%$

| **FRACTION** $\frac{5}{8}$ | | **PERCENTAGE** 62.5% | | **DECIMAL** 0.625 |

$\frac{5}{8} = \frac{5}{8} \times 100\% = 62.5\%$ $62.5\% = \frac{62.5}{100} = 0.625$

$$0.625 = \frac{0.625}{1} = \frac{6.25}{10} = \frac{62.5}{100} = \frac{625}{1000} = \frac{5}{8}$$

VAT

Value Added Tax or VAT is charged on goods you buy or any services you receive. Penny wants to buy this radio cassette player. How much will it cost her if VAT is charged at a rate of $17\frac{1}{2}\%$?

$$VAT = 17\frac{1}{2}\% \text{ of } £50$$

$$= \frac{17\frac{1}{2}}{100} \times £50$$

$$= £8.75$$

Therefore total cost

$$= £50 + £8.75$$

$$= £58.75$$

Purchasing on Credit

This is when you pay a deposit on a purchase and then you make a number of repayments spread over a certain period of time.

Example...

Jim is buying a car on credit. How much will it cost him altogether?

£6,000
or 20% Deposit
& 36 monthly
repayments of £160

20% deposit $= 20\%$ of £6000

$$= \frac{20}{100} \times 6000$$

$$= £1200$$

Total cost of repayments $= 36 \times £160$

$$= £5760$$

Total cost to Jim = £1200 + £5760 = £6960

(This is £960 more than the cash price but he has spread the cost over 36 months)

Simple and Compound Interest

Mr and Mrs Smith have just won £10 000 on the lottery. They decide that they want to invest this money for 2 years. Mr Smith wants to invest the money at 6% simple interest while Mrs Smith wants to invest the money at 6% compound interest.

Simple Interest

Interest gained after 1st year

$$= 6\% \text{ of } £10\,000$$

$$= \frac{6}{100} \times £10\,000 = £600$$

> With simple interest the interest gained (£600) is not added to the amount invested so it does not earn interest in the next year

Interest gained after 2nd year

$$= £600 \text{ (same as 1st year)}$$

Total interest gained

$$= £600 + £600 = £1200$$

Compound Interest

Interest gained after 1st year

$$= 6\% \text{ of } £10\,000$$

$$= \frac{6}{100} \times £10\,000 = £600$$

> With compound interest the interest gained (£600) is added to the amount invested so it earns interest in the next year

Amount invested for 2nd year

$$= £10\,000 + £600$$

$$= £10\,600$$

Interest gained after 2nd year

$$= 6\% \text{ of } £10\,600$$

$$= \frac{6}{100} \times £10\,600 = £636$$

Total interest gained

$$= £600 + £636 = £1\,236$$

Everyday Maths 2

Household Bills

More times than not these are worked out using 'common sense'. Here is a homeowner's electricity bill for one quarter.

The number of units used is found by subtracting the two meter readings

Meter Reading				Amount
Present	Previous	Units used	Pence per unit	
25081	24295	786	7.5	58.95
			Quarterly charge	10.45
		Total charged this quarter excluding VAT		69.40
			VAT at 17.5%	12.15
			Total payable	£81.55

786 units x 7.5 pence per unit = £58.95

You have to pay this regardless of how much electricity you use

17.5% of £69.40
$= \frac{17.5}{100}$ x £69.40
= £12.15

£69.40 + £12.15 = £81.55

Understanding Tables and Charts

The secret to understanding tables and charts is to identify the relevant data. Once you've done this all you need to use is a bit of 'common sense'.

Example...

A rail company operating trains out of Petersfield decides to offer cheap 'Off Peak' fares on services that arrive at London Waterloo after 10am. If a customer who usually catches the 0833 service from Petersfield decides to wait for the first Off Peak train, to take advantage of the offer, what will be the difference in his journey time?

The first thing to do is to identify the relevant data. This is in the third and fourth columns of the timetable shown opposite and has been highlighted for this purpose.

Usual journey time   = Arrival Time − Departure Time

Remember to calculate hours & minutes separately

= 0939hrs − 0833hrs

= 1hr 6min

Off Peak journey time = Arrival Time − Departure Time

= 1018hrs − 0901hrs

= 1hr 17min

Difference in time   = 1hr 17min − 1hr 6min

= 11min

Petersfield, Millford, Farncombe, Woking to London Waterloo

Mondays to Fridays

	AN	NW	AN	AN	AN
Petersfield	0752	0811	0833	0901	0928
Liphook					
Haslemere					
Witley					
Milford (Surrey)	0806	0829	0845	0917	0941
Godalming					
Farncombe	0822	0850	0900	0937	0959
Guildford					
Reading					
Woking	0830	0900	0907	0947	1007
Heathrow Airport (T1)					
Clapham Junction					
London Waterloo	0903	0932	0939	1018	1036

Mondays to Fridays

	AN	NW	AN	AN	AN
Petersfield	0949	0956	1019	1049	1055
Liphook					
Haslemere					
Witley					
Milford (Surrey)	1002	1011	1032	1102	1114
Godalming					
Farncombe	1017	1032	1047	1117	1132
Guildford					
Reading					
Woking	1026	1043	1059	1128	1142
Heathrow Airport (T1)					
Clapham Junction					
London Waterloo	1052	1111	1125	1155	1211

Ratio and Proportion 1

What is a Ratio?

A ratio is a comparison between two or more quantities. Here we have two columns of coins. The first column has ten £1 coins and the second column has six 2p coins. To compare the two sets of coins we can say that the ratio of the number of £1 coins to 2p coins is…

10 to 6 or 10 : 6

Ratios can be simplified into their simplest form, just like fractions…

$$÷2 \left(\begin{array}{c} \mathbf{10 : 6} \\ \mathbf{= 5 : 3} \end{array} \right) ÷2$$

A ratio of **5 : 3** means that for every five £1 coins there are three 2p coins. The above ratio can also be written in the form **1 : n** by dividing both numbers in the ratio by 5…

$$÷5 \left(\begin{array}{c} \mathbf{5 : 3} \\ \mathbf{= 1 : 0.6} \end{array} \right) ÷5$$

It could be written in the form **n : 1** by dividing both numbers in the ratio by 3…

$$÷3 \left(\begin{array}{c} \mathbf{5 : 3} \\ \mathbf{= 1.\dot{6} : 1} \end{array} \right) ÷3$$

Example

A bag of carrots weighs 300g and a bag of potatoes 1.5kg. Calculate the ratio of weight of carrots to weight of potatoes.

Both quantities must be in the same units so,
1.5kg = 1.5 x 1000g = 1500g

Ratio of weight of carrots to weight of potatoes is…

$$÷300 \left(\begin{array}{c} \mathbf{300g : 1500g} \\ \mathbf{= 1 : 5} \end{array} \right) ÷300$$

Ratios and Fractions

A ratio can be written as a fraction and vice versa.

Example

If $\frac{2}{5}$ of a class are boys what is the ratio of boys to girls? Give your answer in the form 1 : n.

The ratio of boys to girls is therefore…

$$×5 \left(\begin{array}{c} \mathbf{\frac{2}{5} : \frac{3}{5}} \\ \mathbf{= 2 : 3} \end{array} \right) ×5$$

Written in the form **1 : n** it is…

$$÷2 \left(\begin{array}{c} \mathbf{2 : 3} \\ \mathbf{= 1 : 1.5} \end{array} \right) ÷2$$

In other words for any one boy in the class there are one and a half girls!

Ratio and Proportion 2

Dividing a Quantity in a Given Ratio

Examples...

1

£60 is to be divided between Jon and Pat in the ratio 2 : 3. How much money does each one receive?

> **We need to divide £60 in the ratio 2 : 3**

The digits in the ratio represent parts. Jon gets 2 parts and Pat gets 3 parts. The total number of parts is **2 + 3 = 5 parts** which is equal to £60. Therefore...

$÷5$ (**5 parts = £60**) $÷5$
(**1 part = £12**)

Since we now know the 'value' of 1 part we can work out how much money Jon and Pat get.

Jon gets 2 parts = 2 x £12 = £24
Pat gets 3 parts = 3 x £12 = £36

> **Check: £24 + £36 = £60**

2 Three brothers aged 6, 9 and 15 decide to share a tin of toffees in the ratio of their ages. If the tin contains 240 toffees how many toffees does each brother get?

> **We need to divide 240 toffees in the ratio 6 : 9 : 15**

> **Whenever possible cancel down your ratio to make things simpler**

$÷3$ (**6 : 9 : 15**
= 2 : 3 : 5) $÷3$

Total number of parts
= 2 + 3 + 5 = 10 parts

Therefore...

$÷10$ (**10 parts = 240 toffees**
1 part = 24 toffees) $÷10$

Brother aged 6 gets 2 x 24 = 48 toffees
Brother aged 9 gets 3 x 24 = 72 toffees
Brother aged 15 gets 5 x 24 = 120 toffees

> **Check: 48 + 72 + 120 = 240 toffees**

Increasing and Decreasing a Quantity in Direct Proportion

Example...

A recipe to make 10 flapjack cakes requires, among other ingredients, 180g of butter. How much butter does a cook need to use if he/she wants to make...

a) 6 flapjack cakes?
b) 25 flapjack cakes?

This is an example of a quantity (e.g. butter) that increases or decreases in direct proportion to the amount of cakes needed. The more cakes that are needed the greater the amount of butter needed and vice versa.

> **The easiest way is to work out the amount of butter needed to make 1 flapjack cake.**

$÷10$ (**10 cakes require 180g of butter**
1 cake requires 18g of butter) $÷10$

a) **6 flapjack cakes require**
6 x 18g = 108g of butter

b) **25 flapjack cakes require**
25 x 18g = 450g of butter

Ratio and Proportion 3

Direct Proportion

Two quantities are in **direct proportion** if when we double (or treble) one quantity the other quantity also doubles (or trebles).

Let us take two quantities, **y** and **x**, that are directly proportional. We can say that…

y is directly proportional to x or y ∝ x,

which means that…

y = kx where **k** is a constant,
i.e. a number that doesn't change.

Whenever you answer questions involving two quantities that are directly proportional, the first thing to do is find the value of **k** to give you the formula for the relationship between the two quantities.

Inverse Proportion

Two quantities are in **inverse proportion** if when we double (or treble) one of the quantities the other quantity halves (or is divided by 3).

Let us take two quantities, **y** and **x**, that are inversely proportional. We can say that…

y is inversely proportional to x or y ∝ $\frac{1}{x}$,

which means that…

y = $\frac{k}{x}$ where k is a constant

Yet again, whenever you answer questions involving two quantities that are inversely proportional, the first thing to do is find the value of **k** to give the formula for the relationship between the two quantities.

> ### Example…
>
> The distance, **d**, travelled by a cyclist is directly proportional to **t**, the time of travel. If the cyclist travels 40 metres in 5 seconds, how far will the same cyclist travel in 12 seconds?
>
> We know that **d ∝ t** or **d = kt**. The first thing we do is find the value for **k**. Since **d = 40m** when **t = 5s** this gives us…
>
> $$40 = k \times 5$$
> $$\therefore k = \frac{40}{5} = 8$$
>
> Our relationship now becomes…
>
> **d = 8 x t or d = 8t**
>
> We can now work out **d** when **t = 12s**
>
> **d = 8t = 8 x 12 = 96m**

> ### Example…
>
> The current, **I**, passing through an electrical component is inversely proportional to the resistance, **R**, of the component. If the current is 0.4 amps when the resistance is 20 ohms, calculate the current when the resistance is 50 ohms.
>
> We know that **I ∝ $\frac{1}{R}$** or **I = $\frac{k}{R}$**. Again, the first thing we do is find the value of **k**. Since **I = 0.4 amps** when **R = 20 ohms**, this gives us…
>
> $$0.4 = \frac{k}{20}$$
> $$\therefore k = 0.4 \times 20 = 8$$
>
> Our relationship now becomes…
>
> **I = $\frac{8}{R}$**
>
> So, we can now work out **I** when **R = 50 ohms**.
>
> **I = $\frac{8}{R} = \frac{8}{50} = 0.16$ amps**

Exponential Growth and Decay

Exponential Growth

When a quantity is increased by a multiplier that is greater than 1 (and for our purpose is constant) over fixed periods of time then we have exponential growth. For example...

A local council recycled 10 000 tonnes of household waste in 2003. Its target is to increase this amount by an extra 20% year on year. Draw a graph to show the recycled waste over the next three years.

We know that the council needs to recycle 20% more compared to the previous year. If the previous year's total is always 100% then the target for the year is 100% + 20%, which is 120%.

$$120\% = \frac{120}{100} = 1.2$$

In other words the amount recycled needs to be multiplied by 1.2 each year. The formula that links the target amount of recycled waste to the amount of recycled waste in 2003 is...

$$A = 10\ 000 \times 1.2^t$$

where **A = target amount of recycled waste** and **t = number of years**

If we draw a table for our results:

Year	2003 (t=0)	2004 (t=1)	2005 (t=2)	2006 (t=3)
A = 10 000 x 1.2^t (tonnes)	10 000	12 000	14 400	17 280

We can now draw our graph:

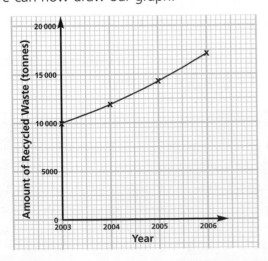

Exponential Decay

When a quantity decays (or decreases) by a multiplier that is less than 1 (and for our purpose is constant) over fixed periods of time then we have exponential decay.

Example...

The graph below, which shows exponential decay, is known to fit the relationship $y = ab^x$. Use the graph to find the value of **a** and **b** and the relationship.

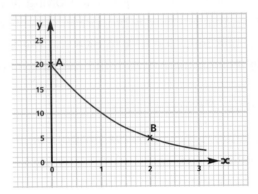

To find **a** and **b** we need to take values for x and **y** from our graph and then substitute them into the relationship $y = ab^x$.

At Point A:

$x = 0$ (curve crosses the y-axis) and **y = 20**
And so... $y = ab^x$
$$20 = ab^0$$
$$20 = a \times 1$$
$$\therefore a = 20$$

Anything to the power of 0 = 1

At Point B:

$x = 2$ and **y = 5** (we know that **a = 20**)
And so... $y = ab^x$
$$5 = 20 \times b^2$$
$$0.25 = b^2$$
$$\therefore b = \sqrt{0.25} = 0.5$$

The relationship is $y = 20 \times 0.5^x$.

The Basics of Algebra 1

Algebra is a branch of mathematics where letters and other symbols are used to represent numbers and quantities in expressions, equations, identities and formulae. Algebra follows the same rules as arithmetic.

Algebraic Expressions

An algebraic expression is a collection of connected letters, numbers and arithmetical symbols. Here are some simple expressions and their meanings.

Algebraic Expression	What it means
$2a$	$a + a$ or $2 \times a$
ab	$a \times b$ (or $b \times a$)
$\dfrac{a}{b}$	$a \div b$
$3a - b$	$(3 \times a) - b$
c^2	$c \times c$
$4mn$	$4 \times m \times n$
x^2	$x \times x$
a^3	$a \times a \times a$
$4x^2y$	$4 \times x \times x \times y$
$(4a)^2$	$4a \times 4a$

Collecting Like Terms

Many expressions can be simplified by collecting together like terms.

Examples...

1 $a + 2a$ is $3a$

2 $5x - 8x + 7x$ is $4x$

When your expression contains 'different' terms, rearrange and collect together all like terms before you simplify.

3 $6b + 3c - 4b$

$= \underline{6b - 4b} + 3c$
　　Like terms

$= 2b + 3c$

4 $4x + 7y - x - 3y$

$= \underline{4x - x} + \underline{7y - 3y}$
　Like terms　　Like terms

$= 3x + 4y$

5 $8x^2 - 4x + 3 - 3x^2 + 7x - 8$

$= \underline{8x^2 - 3x^2} - \underline{4x + 7x} + \underline{3 - 8}$
　　Like terms　　Like terms　Like terms

$= 5x^2 + 3x - 5$

6 $5pq - 7rs + 8qp + 2sr$

$= \underline{5pq + 8pq} - \underline{7rs + 2rs}$

　Like terms　　　Like terms
since 'pq' is the　since 'rs' is the
same as 'qp'　　same as 'sr'

$= 13pq - 5rs.$

Identities

$a + a + a = 3a$ and $a + 2a = 3a$ are both examples of identities and not equations since **3a** is a different way of expressing **a + a + a** or **a + 2a**. In other words, what we have on the left-hand side of the equal sign is no different to what we have on the right-hand side of the equal sign. It is just written in a different way. The unknown (e.g. x) can take any value in an identity and it will always hold true. $5x - 8x + 7x = 4x$ is another example of an identity.

The Basics of Algebra 2

Rules of Indices for Algebra

The same rules are used with letters as we previously used with numbers (see page 15):

1 $x^5 \times x^2$ becomes x^{5+2} and is equal to x^7
… we add the powers.

2 $x^5 \div x^2$ becomes x^{5-2} and is equal to x^3
… we subtract the powers.

3 $(x^5)^2$ becomes $x^{5\times2}$ and is equal to x^{10}
… we multiply the powers.

Also…

4 x^1 is the same as x … and vice versa
… x is the same as x^1.

5 x^0 is equal to 1 … anything to the power of 0 (zero) is always equal to 1.

6 x^{-1} is the same as $\frac{1}{x}$ … and vice versa
… $\frac{1}{x}$ is the same as x^{-1}.

Examples…

1 $2x^3 \times 4x^2$
$= (2 \times 4)x^{3+2}$
$= 8x^5$

> Multiply your numbers as normal and add the powers

2 $(3x^4)^2$
$= 3^2 x^{4\times2}$
$= 9x^8$

> Square your number as normal and then multiply the powers of x

3 $\dfrac{10s^5}{5s^2}$
$= 2s^{5-2}$
$= 2s^3$

> Divide your numbers, then subtract the powers of s

4 $(3pq^2)^3$
$= 3^3 p^{1\times3} q^{2\times3}$
$= 27p^3 q^6$

> Cube your number as normal and multiply the powers of p and q separately

5 $(x^{-3})^{-2}$
$= x^{-3\times-2}$
$= x^6$

> Multiply the powers of x

6 $\dfrac{5r^3 \times 4r^2}{10r^6}$
$= \dfrac{(5 \times 4)r^{3+2}}{10r^6}$
$= \left(\dfrac{20}{10}\right)r^{5-6}$
$= 2r^{-1} \left(= \dfrac{2}{r}\right)$

> Do one operation at a time. Multiplication first…

> … and then division

7 $\dfrac{s^4 r^5 \times s^{-2} r^{-5}}{s^2}$
$= \dfrac{s^{4-2} r^{5-5}}{s^2}$
$= \dfrac{s^2 r^0}{s^2}$
$= s^0$
$= 1$

> Do one operation at a time. Add the powers of s and r respectively…

> … and then subtract the powers of s. Remember, anything to the power of 0 (zero) is equal to 1

This involves substituting numbers for letters in expressions. The simplest substitution would involve only one operation (e.g. an addition or multiplication). However, when a substitution involves more than one operation, you must perform them in the order shown below.

Also, since you are expected to be able to substitute positive and negative numbers into expressions turn back to page 10 to refresh yourself on the multiplication and division of integers.

B I D M A S

| BRACKETS | INDICES | DIVISIONS AND MULTIPLICATIONS - THESE CAN BE DONE IN ANY ORDER | ADDITIONS AND SUBTRACTIONS - AGAIN IN ANY ORDER |

Examples...

If **a = 3, b = 8, c = 20, d = -4,** calculate the value of…

1 $2(a + b) - c$ — Substitute in your numbers

$= 2(3 + 8) - 20$ — Work out the bracket first

$= 2 \times 11 - 20$ — Then the multiplication

$= 22 - 20$ — Then the subtraction

$= 2$

2 $3d^2 + c$ — Substitute in your numbers

$= 3 \times (-4)^2 + 20$ — Work out the square first ($- \times - = +$)

$= 3 \times 16 + 20$ — Then the multiplication

$= 48 + 20$ — Then the addition

$= 68$

3 $\dfrac{c(d + 1)}{5}$ — Substitute in your numbers

$= \dfrac{20(-4 + 1)}{5}$ — Work out the bracket first

$= \dfrac{20 \times -3}{5}$ — Then the multiplication ($+ \times - = -$)

$= \dfrac{-60}{5}$ — Then the division ($- \div + = -$)

$= -12$

4 $\dfrac{1}{2} d^3$ — Substitute in your numbers

$= \dfrac{1}{2} \times (-4)^3$ — Work out the cube first ($- \times - = +$ and then $+ \times - = -$)

$= \dfrac{1}{2} \times -64$ — Then the multiplication ($+ \times - = -$)

$= -32$

Brackets and Factorisation

Multiplying Out Brackets

When we multiply out a bracket, everything which is inside the bracket must be multiplied by whatever is immediately outside the bracket.

Examples...

1 $4(x + 3)$

$= 4 \times x + 4 \times 3$

$= 4x + 12$

2 $3x(4x^2 - 5)$

$= 3x \times 4x^2 + 3x \times -5$

$= 12x^3 - 15x$

If your expression includes two brackets, it may result in like terms, which you need to collect together and simplify.

3 $2(4x + 3) + 5(x - 2)$

$= 2 \times 4x + 2 \times 3 + 5 \times x + 5 \times -2$

$= 8x + 6 + 5x - 10$

$= \underbrace{8x + 5x}_{\text{Like terms}} + \underbrace{6 - 10}_{\text{Like terms}}$

$= 13x - 4$

4 $x(5 - x) + 4x(2x^2 + 3x)$

$= x \times 5 + x \times -x + 4x \times 2x^2 + 4x \times 3x$

$= 5x - x^2 + 8x^3 + 12x^2$

$= 8x^3 \underbrace{- x^2 + 12x^2}_{\text{Like terms}} + 5x$

$= 8x^3 + 11x^2 + 5x$

When you multiply out two brackets make sure that each term in the second bracket is multiplied by each term in the first bracket.

5 $(x + 2)(x + 3)$

$= x(x + 3) + 2(x + 3)$

$= x \times x + x \times 3 + 2 \times x + 2 \times 3$

$= x^2 + \underbrace{3x + 2x}_{\text{Like terms}} + 6$

$= x^2 + 5x + 6$

6 $(3x - 2)^2$

$= (3x - 2)(3x - 2)$

$= 3x(3x - 2) - 2(3x - 2)$

$= 3x \times 3x + 3x \times -2 - 2 \times 3x - 2 \times -2$

$= 9x^2 \underbrace{- 6x - 6x}_{\text{Like terms}} + 4$

$= 9x^2 - 12x + 4$

Factorisation

This is the reverse process to multiplying out brackets. An expression is rewritten with a bracket by taking out highest common factors.

Examples...

1 $4x + 6 = 2(2x + 3)$ ... as **2** is the highest common factor of both **4** and **6**

2 $6x^2 + 8x = 2x(3x + 4)$ ... as **2x** is the highest common factor of both **6x²** and **8x**

3 $9x^3y^2 - 6xy^3 = 3xy^2(3x^2 - 2y)$ ... as **3xy²** is the highest common factor of both **9x³y²** and **6xy³**

4 $x(2y + 3) + z(2y + 3) = (2y + 3)(x + z)$ ... as **(2y + 3)** is a common factor of **x(2y + 3)** and **z(2y + 3)**

To check each of the examples above multiply out each bracket. You should end up with the original expression.

Solving Linear Equations 1

Equations such as...

$$4x = 12$$
$$x + 3 = 7$$
$$2(x + 5) = 14$$

... are all examples of linear equations since the highest power they contain is x^1 (i.e. x).

Each of these linear equations can be solved to find the 'unknown' value of x by collecting all the x's on one side and all the numbers on the other side. The simplest linear equation would involve one operation to solve it. Most however require at least two operations.

Examples...

1

$$6x = 15$$
$$\frac{6x}{6} = \frac{15}{6}$$
$$x = 2.5$$

- Divide both sides of the equation by **6** to leave just x on the left-hand side.

2

$$3x - 5 = 19$$
$$3x - 5 + 5 = 19 + 5$$
$$\frac{3x}{3} = \frac{24}{3}$$
$$x = 8$$

- Add **5** to both sides of the equation to remove the **-5** from the left-hand side.
- Divide both sides of the equation by **3** to leave just x on the left-hand side.

3

$$18 = 4(x + 9)$$
$$18 = 4x + 36$$
$$18 - 36 = 4x + 36 - 36$$
$$\frac{-18}{4} = \frac{4x}{4}$$
$$-4.5 = x$$
$$\text{or } x = -4.5$$

- Multiply out the bracket on the right-hand side.
- Subtract **36** from both sides of the equation to remove the **+36** on the right-hand side.
- Divide both sides of the equation by **4** to leave just x on the right-hand side.

4

$$\frac{4x + 3}{3} + \frac{5x - 3}{4} = 8$$
$$\frac{^{4}12 \times (4x + 3)}{3} + \frac{^{3}12 \times (5x - 3)}{4} = 12 \times 8$$
$$16x + 12 + 15x - 9 = 96$$
$$31x + 3 = 96$$
$$31x + 3 - 3 = 96 - 3$$
$$\frac{31x}{31} = \frac{93}{31}$$
$$x = 3$$

- Multiply all terms on both sides by **12**, which is the lowest common multiple of **3** and **4**, to remove the fractional values.
- Collect like terms on the left-hand side of the equation.
- Subtract **3** from both sides of the equation to remove the **+3** on the left-hand side.
- Divide both sides of the equation by **31** to leave just x on the left-hand side.

Solving Linear Equations 2

The following examples have the 'unknown', e.g. x, appearing on both sides of the equation. Once again they are solved by collecting all the x's on one side and all the numbers on the other side.

It does not really matter on which side of the equal sign you collect all the x's, it is your choice. In the examples below, they are collected on the side that has the most positive x's to start with. This ensures that you end up with a value for x rather than $-x$.

Examples...

1

$$8x - 7 = 5x + 26$$
$$8x - 5x - 7 = 5x - 5x + 26$$
$$3x - 7 + 7 = 26 + 7$$
$$\frac{3x}{3} = \frac{33}{3}$$
$$x = 11$$

- Subtract **5x** from both sides of the equation to leave all the x's on the left-hand side.
- Add **7** to both sides of the equation to remove the **-7** on the left-hand side.
- Divide both sides of the equation by **3** to leave just x on the left-hand side.

2

$$4(3x - 2) = 15x + 22$$
$$12x - 8 = 15x + 22$$
$$12x - 12x - 8 = 15x - 12x + 22$$
$$-8 - 22 = 3x + 22 - 22$$
$$\frac{-30}{3} = \frac{3x}{3}$$
$$-10 = x$$
$$\text{or } x = -10$$

- Multiply out the bracket on the left-hand side.
- Subtract **12x** from both sides of the equation to leave all the x's on the right-hand side.
- Subtract **22** from both sides of the equation to remove the **22** on the right-hand side.
- Divide both sides of the equation by **3** to leave just x on the right-hand side.

3

$$\frac{19 + x}{2} = 2 - x$$
$$2 \times \frac{(19 + x)}{2} = 2 \times (2 - x)$$
$$19 + x = 4 - 2x$$
$$19 + x + 2x = 4 - 2x + 2x$$
$$19 - 19 + 3x = 4 - 19$$
$$\frac{3x}{3} = \frac{-15}{3}$$
$$x = -5$$

- Multiply all terms on both sides by **2** to remove the fractional value.
- Add **2x** to both sides of the equation to remove the **-2x** on the right-hand side.
- Subtract **19** from both sides of the equation to remove the **19** on the left-hand side.
- Divide both sides of the equation by **3** to leave just x on the left-hand side.

Problem Solving Using Linear Equations

This involves being given information, forming a linear equation from the information given and then solving the equation.

Example...

The four angles of a quadrilateral are: **a, a + 20°, a + 40° and a + 60°**. Calculate the size of each angle.

Firstly, we know that the angles of a quadrilateral add up to **360°**. Therefore...

$$a + (a + 20) + (a + 40) + (a + 60) = 360$$
$$4a + 120 = 360$$
$$4a = 240 \quad \longleftarrow \text{Subtract 120 from both sides}$$
$$a = 60° \quad \longleftarrow \text{Divide both sides by 4}$$

$a = 60°$, $a + 20° = 80°$, $a + 40° = 100°$, $a + 60° = 120°$.

Formulae 1

Formulae

Formulae show the relationship between two or more changeable quantities (variables). They can be written in words, but most often symbols are used instead. For example, the area of a circle is given by the formula: $A = \pi r^2$, which describes the relationship between the area of a circle and its radius.

Formulae can be rearranged to make a different letter the 'subject', e.g. $a = b + c$ has **a** as the subject since it is on one side by itself. If we wanted to make **b** or **c** the subject then we would need to rearrange the formula by moving terms from one side of the equals sign to the other.

Examples...

1 Make **b** the subject of the following formula.

$$a = b + c$$
$$a - c = b + \cancel{c} - \cancel{c}$$
or $b = a - c$

- SUBTRACT **c** from both sides of the formula. This will remove the **+c** on the right-hand side to leave **b** on its own.
- REWRITE the formula with **b** on the left-hand side.

---

2 Make **x** the subject of the following formula.

$$4(3x + 2y) = 5x + 2z$$
$$12x + 8y = 5x + 2z$$
$$12x - 5x + 8y = \cancel{5x} - \cancel{5x} + 2z$$
$$7x + \cancel{8y} - \cancel{8y} = 2z - 8y$$
$$\frac{\cancel{7}x}{\cancel{7}} = \frac{2z - 8y}{7}$$
$$x = \frac{2z - 8y}{7}$$

- MULTIPLY out the bracket on the left-hand side.
- SUBTRACT **5x** from both sides of the formula to remove the **5x** on the right-hand side.
- SUBTRACT **8y** from both sides of the formula to remove the **+8y** on the left-hand side.
- DIVIDE both sides of the formula by **7** to leave just **x** on the left-hand side.

---

3 Make **g** the subject of the following formula.

$$e = f + \frac{g}{d}$$
$$d \times e = d \times f + \cancel{d} \times \frac{g}{\cancel{d}}$$
$$de - df = \cancel{df} - \cancel{df} + g$$
$$g = de - df$$

- MULTIPLY all terms on both sides by **d** to remove the fractional value.
- SUBTRACT **df** from both sides of the formula to leave just **g** on the right-hand side.
- REWRITE the formula with **g** on the left-hand side.

---

4 Make **r** the subject of the following formula.

$$A = \pi r^2$$
$$\frac{A}{\pi} = \frac{\cancel{\pi} r^2}{\cancel{\pi}}$$
$$\sqrt{\frac{A}{\pi}} = \sqrt{r^{\cancel{2}}}$$
$$\sqrt{\frac{A}{\pi}} = r \text{ or } r = \sqrt{\frac{A}{\pi}}$$

- DIVIDE both sides of the formula by π to remove the π on the right-hand side.
- Take the SQUARE ROOT of both sides of the formula to remove the 'square' and leave just **r** on the right-hand side.
- REWRITE the formula with **r** on the left-hand side.

Formulae 2

Using Formulae

Whenever you use a formula to work out an unknown quantity all you are doing is substituting numbers for symbols.

Examples...

1 The formula that converts a temperature reading from degrees Celsius (°C) into degrees Fahrenheit (°F) is: $F = \frac{9}{5}C + 32$
What is the temperature in degrees Fahrenheit if the temperature in degrees Celsius is 25°C?

Our formula is: $F = \frac{9}{5}C + 32$

$F = \frac{9}{5} \times 25 + 32$

$F = 45 + 32$

$F = 77°F$

> We now substitute into our formula a value for **C**, which is 25°C, to work out **F**

---

2 The area of a triangle is given by the formula:

$$A = \frac{B \times H}{2}$$

where **B** is length of base and **H** is height of triangle. Calculate the height of a triangle of area 30cm² if the length of its base is 10cm.

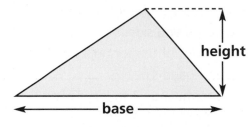

Our formula is:

$$A = \frac{B \times H}{2}$$

> Since we want the subject of the formula to be **H**, we need to rearrange.

$$A \times 2 = \frac{B \times H}{2} \times 2$$

> Multiply both sides by 2

$$\frac{2A}{B} = \frac{B \times H}{B}$$

> Divide both sides by **B** to leave **H** on its own

$$H = \frac{2A}{B}$$

$$H = \frac{2 \times 30}{10} = \frac{60}{10}$$

> We can now substitute in values for **A** and **B** to find **H**

$$H = 6cm$$

Generating Formulae

This means writing a formula using given information. You may then need to use this formula to work out an unknown quantity. For example...

Generate a formula for the perimeter of a rectangle in terms of its area, **A**, and width, **w**, only. Use it to work out the perimeter if area = 40cm² and width = 4cm

$$A = \ell \times w$$

$$\text{Perimeter} = 2\ell + 2w$$

width (w) • Length (ℓ)

Firstly, we use the formula for area and rearrange it to give us ℓ on its own...

$$A = \ell \times w \longrightarrow \ell = \frac{A}{w}$$

Perimeter $= 2\ell + 2w$

> ... and then substitute for ℓ in our formula for perimeter.

$$= 2\frac{A}{w} + 2w$$

$$= \frac{2A}{w} + 2w$$

> We now have a formula for perimeter in terms of area and width only. Therefore we can substitute in values for **A** and **w** to find the perimeter.

$$= \frac{2 \times 40}{4} + 2 \times 4$$

$$= \frac{80}{4} + 8$$

$$= 20 + 8 = 28cm$$

Quadratic Expressions 1

A quadratic expression is an expression in which the highest power of x is x^2, e.g. $x^2 + 6x + 5$, $x^2 - 9$ or $x^2 - 3x$. Quadratic expressions with terms that have no common factors give us two brackets when factorised, where each bracket contains a letter (usually x) and a number. For example, to factorise a quadratic expression such as $x^2 + 6x + 5$ you must always follow these three steps:

1 Write out two brackets and put an x in each one… $(x \qquad)(x \qquad)$

2 Find two numbers which multiply to give the end number $+5$ ($+1 \times +5 = +5$) and add up to give the middle number $+6$ ($+1 + +5 = +6$)

3 Insert your numbers - one in each bracket. Therefore, $x^2 + 6x + 5 = (x + 1)(x + 5)$

Here are three other examples…

---

1 $x^2 + 4x + 3$
$= (x + 1)(x + 3)$

Check:
$(x + 1)(x + 3)$
$= x^2 + 3x + x + 3$
$= x^2 + 4x + 3$ ✓

- Write out your two brackets and put an x in each one.
- We now need two numbers that multiply to give $+3$ and also add up to $+4$.

$+1 \times +3 = +3$ ✓ and $+1 + +3 = +4$ ✓ (these are our two numbers)
$-1 \times -3 = +3$ ✓ and $-1 + -3 = -4$ ✗

---

2 $x^2 - 8x + 7$
$= (x - 1)(x - 7)$

Check:
$(x - 1)(x - 7)$
$= x^2 - 7x - 1x + 7$
$= x^2 - 8x + 7$ ✓

- Write out your two brackets and put an x in each one.
- We now need two numbers that multiply to give $+7$ and also add up to -8.

$+1 \times +7 = +7$ ✓ and $+1 + +7 = +8$ ✗
$-1 \times -7 = +7$ ✓ and $-1 + -7 = -8$ ✓ (these are our two numbers)

---

3 $x^2 + 5x - 6$
$= (x - 1)(x + 6)$

Check:
$(x - 1)(x + 6)$
$= x^2 + 6x - x - 6$
$= x^2 + 5x - 6$ ✓

- Write out your two brackets and put an x in each one.
- We now need two numbers that multiply to give -6 and also add up to $+5$.

$+1 \times -6 = -6$ ✓ and $+1 + -6 = -5$ ✗
$-1 \times +6 = -6$ ✓ and $-1 + +6 = +5$ ✓ (these are our two numbers)
$+2 \times -3 = -6$ ✓ and $+2 + -3 = -1$ ✗
$-2 \times +3 = -6$ ✓ and $-2 + +3 = +1$ ✗

Difference of Two Squares

There are some quadratic expressions that don't have any x's in the middle and end in – 'a number'^{2}. These expressions can also be factorised. Expressions such as $x^2 - 1^2$, $x^2 - 2^2$, $x^2 - 3^2$, $x^2 - 4^2$ and so on, are known as the 'difference of two squares'. Again, you should check your answers by multiplying out the brackets. If you do so you should end up with the original expression.

Examples…

1 $x^2 - 1 = x^2 - 1^2 = (x + 1)(x - 1)$

2 $x^2 - 4 = x^2 - 2^2 = (x + 2)(x - 2)$

3 $x^2 - 9 = x^2 - 3^2 = (x + 3)(x - 3)$

4 $x^2 - 16 = x^2 - 4^2 = (x + 4)(x - 4)$

Quadratic Expressions 2

On the previous page you factorised quadratic expressions where the number of x^2s in the x^2 term was 1. You now need to move one step further and factorise quadratic expressions such as $2x^2 + 7x + 5$ and $5x^2 + 6x - 8$, where they all have a number greater than 1 in front of the x^2 term. For example, to factorise a quadratic expression such as $2x^2 + 7x + 5$ you must always follow these three steps:

1 Write out two brackets and put an x in one and $2x$ in the other... $(x \quad)(2x \quad)$. These two terms, x and $2x$, when multiplied together must equal $2x^2$...the x^2 term in our expression.

2 Find two numbers which multiply to give the end number +5 (+1 x +5 = +5) and when multiplied by the x and $2x$ terms add up to give us the middle term +7x [(5 x x) + (1 x 2x) = +7x].

3 Insert your numbers - one in each bracket where the two terms +5 and x and the two terms +1 and 2x must be in **opposite brackets**. Therefore, $2x^2 + 7x + 5 = (x + 1)(2x + 5)$.

Here are two other examples...

---

1 $\quad 3x^2 - 8x + 5$

$= (x - 1)(3x - 5)$

Check:

$(x - 1)(3x - 5)$

$= 3x^2 - 5x - 3x + 5$

$= 3x^2 - 8x + 5$ ✓

- Write out your two brackets and put an x in one and a $3x$ in the other, since $x \times 3x = 3x^2$
- We now need two numbers that multiply to give **+5** and when multiplied by x and $3x$ add up to **-8**x.

+1 x +5 = +5 ✓ and	(+1 x x) + (+5 x 3x) = +16x ✗
+1 x +5 = +5 ✓ and	(+1 x 3x) + (+5 x x) = +8x ✗
-1 x -5 = +5 ✓ and	(-1 x x) + (-5 x 3x) = -16x ✗
-1 x -5 = +5 ✓ and	(-1 x 3x) + (-5 x x) = -8x ✓

- **Remember...** The two terms **-1** and **3x** and the two terms **-5** and x must be in opposite brackets.

---

2 $\quad 5x^2 + 9x - 2$

$= (x + 2)(5x - 1)$

Check:

$(x + 2)(5x - 1)$

$= 5x^2 - x + 10x - 2$

$= 5x^2 + 9x - 2$ ✓

- Write out your two brackets and put an x in one and a $5x$ in the other, since $x \times 5x = 5x^2$
- We now need two numbers that multiply to give **-2** and when multiplied by x and $5x$ add up to **+9x**.

+1 x -2 = -2 ✓ and	(+1 x x) + (-2 x 5x) = -9x ✗
+1 x -2 = -2 ✓ and	(+1 x 5x) + (-2 x x) = +3x ✗
-1 x +2 = -2 ✓ and	(-1 x x) + (+2 x 5x) = +9x ✓
-1 x +2 = -2 ✓ and	(-1 x 5x) + (+2 x x) = -3x ✗

- **Remember...** The two terms **-1** and x and the two terms **+2** and 5x must be in opposite brackets.

Quadratic Equations 1

Solving Quadratic Equations by Factorisation

Equations such as…

$x^2 + 4x + 3 = 0$

$x^2 - 16 = 0$

$2x^2 + 5x - 3 = 0$

… are all examples of quadratic equations (i.e. their highest power is x^2).

To solve any quadratic equation there are two steps you must follow:

1 Factorise the quadratic expression within the equation to form two brackets (see pages 43 & 44).

2 Treat each bracket as a separate linear equation equal to 0 (zero).

Examples…

1 $x^2 - 6x + 5 = 0$

Our factorised quadratic equation now looks like…

$(x - 1)(x - 5) = 0$

To factorise the quadratic expression within the equation, we need two numbers that multiply to give **+5** and also add up to **-6**.

+1 x +5 = +5 ✓ and +1 + +5 = +6 ✗

-1 x -5 = +5 ✓ and -1 + -5 = -6 ✓ (these are our two numbers)

- We now have TWO BRACKETS WHICH WHEN MULTIPLIED TOGETHER ARE EQUAL TO 0 (zero).

- This is only possible if ONE OR BOTH OF THE TWO BRACKETS IS ALSO EQUAL TO 0 (zero).

- We now TREAT EACH BRACKET AS A SEPARATE LINEAR EQUATION EQUAL TO 0 (zero).

And so…

$(x - 1) = 0$

$x - \cancel{1} + \cancel{1} = 0 + 1$

$x = 1$ … is one solution

Check: $x = 1$

$x^2 - 6x + 5 = 0$

$1^2 - 6 \times 1 + 5 = 0$

$1 - 6 + 5 = 0$ ✓

or

$(x - 5) = 0$

$x - \cancel{5} + \cancel{5} = 0 + 5$

$x = 5$ …is the other

Check: $x = 5$

$x^2 - 6x + 5 = 0$

$5^2 - 6 \times 5 + 5 = 0$

$25 - 30 + 5 = 0$ ✓

2 $x^2 - 5x - 14 = 0$

Our factorised quadratic equation now looks like…

$(x + 2)(x - 7) = 0$

To factorise the quadratic expression within the equation, we need two numbers that multiply to give **-14** and also add up to **-5**.

+1 x -14 = -14 ✓ and +1 + -14 = -13 ✗

-1 x +14 = -14 ✓ and -1 + +14 = +13 ✗

+2 x -7 = -14 ✓ and +2 + -7 = -5 ✓ (these are our two numbers)

-2 x +7 = -14 ✓ and -2 + +7 = +5 ✗

And so…

$(x + 2) = 0$

$x + \cancel{2} - \cancel{2} = 0 - 2$

$x = -2$ … is one solution

Check: $x = -2$

$x^2 - 5x - 14 = 0$

$(-2)^2 - 5 \times -2 - 14 = 0$

$4 + 10 - 14 = 0$ ✓

or

$(x - 7) = 0$

$x - \cancel{7} + \cancel{7} = 0 + 7$

$x = 7$ … is the other

Check: $x = 7$

$x^2 - 5x - 14 = 0$

$7^2 - 5 \times 7 - 14 = 0$

$49 - 35 - 14 = 0$ ✓

Quadratic Equations 2

Solving Quadratic Equations by Completing the Square

This method can also be used to solve quadratic equations and is particularly useful when the solutions are not integers. Although you can use this method you should always try and factorise the quadratic expression within the quadratic equation first (as on page 45) in order to solve the equation.

1 Rearrange your equation so all the terms involving x^2 and x are on one side and your number is on the other side.

2 Write the terms in the form… $(x + a \text{ 'number'})^2$ Your 'number' should have a value **half** that of the number before the term involving x.

3 Now expand the square bracket to get the original terms involving x^2 and x, **plus a new 'number'**. To balance the equation, add the new 'number' onto the other side too.

4 Take the square root of both sides of your equation to get two separate linear equations which you then solve.

Examples…

1 $x^2 - 6x + 5 = 0$

$x^2 - 6x = -5$

$(x - 3)^2 = -5 + 9$

$(x - 3)^2 = 4$

$\sqrt{(x - 3)^2} = \sqrt{4}$

$x - 3 = \pm 2$

- Rearrange your equation.
- Complete the square and balance the equation by adding **+9** to the right-hand side.
  $[(x - 3)^2 = (x - 3)(x - 3) = x^2 - 6x + 9]$
- Take the square root of both sides. Remember that $\sqrt{4} = \pm 2$ (see page 16). We now have two separate linear equations: $x - 3 = +2$ and $x - 3 = -2$.

And so…

$x - 3 = +2$

$x = 5$

… is one solution

Check:

$x^2 - 6x + 5 = 0$

$5^2 - 6 \times 5 + 5 = 0$

$25 - 30 + 5 = 0$ ✓

or

$x - 3 = -2$

$x = 1$

… is the other

Check:

$x^2 - 6x + 5 = 0$

$1^2 - 6 \times 1 + 5 = 0$

$1 - 6 + 5 = 0$ ✓

2 $x^2 - 4x - 6 = 0$

$x^2 - 4x = 6$

$(x - 2)^2 = 6 + 4$

$(x - 2)^2 = 10$

$\sqrt{(x - 2)^2} = \sqrt{10}$

$x - 2 = \pm 3.16$

- Rearrange your equation.
- Complete the square and balance the equation by adding **+4** to the right-hand side.
  $[(x - 2)^2 = (x - 2)(x - 2) = x^2 - 4x + 4]$
- Take the square root of both sides. Remember that $\sqrt{10} = \pm 3.16$ (to 2d.p.). We now have two separate linear equations: $x - 2 = +3.16$ and $x - 2 = -3.16$.

And so…

$x - 2 = +3.16$

$x = 5.16$

… is one solution

Check:

$x^2 - 4x - 6 = 0$

$(5.16)^2 - (4 \times 5.16) - 6 = 0$

$26.63 - 20.64 - 6 = 0$ ✓

or

$x - 2 = -3.16$

$x = -1.16$

… is the other

Check:

$x^2 - 4x - 6 = 0$

$(-1.16)^2 - (4 \times -1.16) - 6 = 0$

$1.35 + 4.64 - 6 = 0$ ✓

Quadratic Equations 3

Solving Quadratic Equations Using the Quadratic Formula

This formula can be used to solve quadratic equations. This method is again useful when the solutions are not whole numbers.

The formula is $x = \dfrac{-b \pm \sqrt{b^2 - 4ac}}{2a}$

where any quadratic equation can be written in the form: $ax^2 + bx + c = 0$ when

a = number before the term involving x^2
b = number before the term involving x
c = end number

Examples...

1 $x^2 - 6x + 5 = 0$

Therefore, $a = 1$, $b = -6$ and $c = +5$. If we now substitute these values into the formula:

$x = \dfrac{-b \pm \sqrt{b^2 - 4ac}}{2a}$

$x = \dfrac{--6 \pm \sqrt{(-6)^2 - (4 \times 1 \times +5)}}{2 \times 1}$

$x = \dfrac{6 \pm \sqrt{36 - 20}}{2}$

$x = \dfrac{6 \pm \sqrt{16}}{2}$

$x = \dfrac{6 \pm 4}{2}$

We now have two separate linear equations which can be solved.

And so...

$x = \dfrac{6 + 4}{2}$

$x = \dfrac{10}{2}$

$x = 5$

or...

$x = \dfrac{6 - 4}{2}$

$x = \dfrac{2}{2}$

$x = 1$

Check:
$x^2 - 6x + 5 = 0$
$(5)^2 - (6 \times 5) + 5 = 0$
$25 - 30 + 5 = 0$ ✓

Check:
$x^2 - 6x + 5 = 0$
$(1)^2 - (6 \times 1) + 5 = 0$
$1 - 6 + 5 = 0$ ✓

2 $5x^2 + 2x - 4 = 0$

Therefore, $a = 5$, $b = +2$ and $c = -4$. If we now substitute these values into the formula:

$x = \dfrac{-b \pm \sqrt{b^2 - 4ac}}{2a}$

$x = \dfrac{-+2 \pm \sqrt{(+2)^2 - (4 \times 5 \times -4)}}{2 \times 5}$

$x = \dfrac{-2 \pm \sqrt{4 + 80}}{10}$

$x = \dfrac{-2 \pm \sqrt{84}}{10}$

$x = \dfrac{-2 \pm 9.165}{10}$

We now have two separate linear equations which can be solved.

And so...

$x = \dfrac{-2 + 9.165}{10}$

$x = \dfrac{7.165}{10}$

$x = 0.72$ (to 2 d.p.)

or...

$x = \dfrac{-2 - 9.165}{10}$

$x = \dfrac{-11.165}{10}$

$x = -1.12$ (to 2 d.p.)

Check:
$5x^2 + 2x - 4 = 0$
$5(0.72)^2 + (2 \times 0.72) - 4 = 0$
$2.59 + 1.44 - 4 = 0$ ✓

Check:
$5x^2 + 2x - 4 = 0$
$5(-1.12)^2 + (2 \times -1.12) - 4 = 0$
$6.27 - 2.24 - 4 = 0$ ✓

Trial and Improvement

This method can be used to find a solution to any equation. As the name suggests we trial a possible solution by substituting its value into the equation.

The process is then repeated using a different possible solution and so on. The idea is that each subsequent trial is an improvement on the previous trial.

Examples...

1 The equation $x^3 + x = 20$ has a solution somewhere between $x = 2$ and $x = 3$. By trial and improvement calculate a solution to the equation to 2 decimal places.

Since we are told that there is a solution between $x = 2$ and $x = 3$ we will substitute their values into our equation to see what $x^3 + x$ gives us. Remember we want its value to be equal to **20**.

x	$x^3 + x$	Comment
2	$2^3 + 2 = 8 + 2 = \mathbf{10}$	Less than 20
3	$3^3 + 3 = 27 + 3 = \mathbf{30}$	More than 20
Try 2.5	$2.5^3 + 2.5 = 15.625 + 2.5 = \mathbf{18.125}$	Less than 20
Try 2.6	$2.6^3 + 2.6 = 17.576 + 2.6 = \mathbf{20.176}$	More than 20
Try 2.55	$2.55^3 + 2.55 = 16.581375 + 2.55 = \mathbf{19.131375}$	Less than 20
Try 2.58	$2.58^3 + 2.58 = 17.173512 + 2.58 = \mathbf{19.753512}$	Less than 20
Try 2.59	$2.59^3 + 2.59 = 17.373979 + 2.59 = \mathbf{19.963979}$	Less than 20
Try 2.595	$2.595^3 + 2.595 = 17.474795 + 2.595 = \mathbf{20.069795}$	Just more than 20

So $x = \mathbf{2.59}$ (to 2 d.p.) since the last trial above ($x = \mathbf{2.595}$) is more than 20.

If this trial had been less than 20 then $x = \mathbf{2.60}$ (to 2 d.p.) would be the solution. (It is important to try the middle value (**2.595**)).

• •

2 The equation $x^2 - \frac{12}{x} = 10$ has a solution somewhere between $x = 3$ and $x = 4$.

By trial and improvement calculate a solution to the equation to 2 decimal places.

x	$x^2 - \frac{12}{x}$	Comment
3	$3^2 - \frac{12}{3} = 9 - 4 = \mathbf{5}$	Less than 10
4	$4^2 - \frac{12}{4} = 16 - 3 = \mathbf{13}$	More than 10
Try 3.5	$3.5^2 - \frac{12}{3.5} = 12.25 - 3.4285714 = \mathbf{8.8214286}$	Less than 10
Try 3.6	$3.6^2 - \frac{12}{3.6} = 12.96 - 3.3333333 = \mathbf{9.6266667}$	Less than 10
Try 3.7	$3.7^2 - \frac{12}{3.7} = 13.69 - 3.2432432 = \mathbf{10.446757}$	More than 10
Try 3.65	$3.65^2 - \frac{12}{3.65} = 13.3225 - 3.2876712 = \mathbf{10.034829}$	More than 10
Try 3.64	$3.64^2 - \frac{12}{3.64} = 13.2496 - 3.2967033 = \mathbf{9.9528967}$	Less than 10
Try 3.645	$3.645^2 - \frac{12}{3.645} = 13.286025 - 3.2921811 = \mathbf{9.9938439}$	Just less than 10

So $x = \mathbf{3.65}$ (to 2 d.p.) since the last trial above ($x = \mathbf{3.645}$) is less than 10.

If this trial had been more than 10 then $x = \mathbf{3.64}$ (to 2 d.p.) would be the solution. (It is important to try the middle value (**3.645**)).

Number Patterns & Sequences 1

A number pattern or sequence is a series of numbers which follow a rule. Each number in a sequence is called a **term**, where the first number in the sequence is called the **1st term** and so on.

1st term	2nd term	3rd term	4th term		The next two terms

2, 6, 10, 14, ...
+4 +4 +4

The rule is that each term is **4 more** than the previous term. These terms have a common difference of +4

... 18, 22, ...
+4

20, 17, 14, 11, ...
-3 -3 -3

The rule is that each term is **3 less** than the previous term. These terms have a common difference of -3

... 8, 5, ...
-3

1, 3, 9, 27, ...
x3 x3 x3

The rule is that each term is **3 times** the previous term. These terms don't have a common difference between them

... 81, 243, ...
x3

A sequence can also be a series of diagrams.

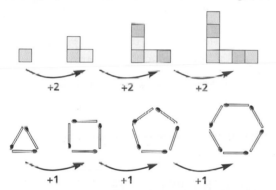

+2 +2 +2

Each diagram (term) has **2 more** boxes in it than the previous diagram. These diagrams have a common difference of +2

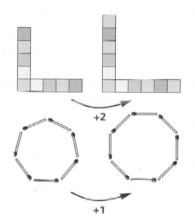

+2

Each diagram (term) has **1 more** match in it than the previous diagram. These diagrams have a common difference of +1

+1 +1 +1

+1

Other examples...

1 Squared integers

1, 4, 9, 16, ...
$(1^2 = 1)$ $(2^2 = 4)$ $(3^2 = 9)$ $(4^2 = 16)$

2 Triangular numbers

1, 3, 6, 10, ...
+2 +3 +4

3 Powers of 2

1, 2, 4, 8, ...
$(2^0 = 1)$ $(2^1 = 2)$ $(2^2 = 4)$ $(2^3 = 8)$

4 Powers of 10

1, 10, 100, 1000, ...
$(10^0 = 1)$ $(10^1 = 10)$ $(10^2 = 100)$ $(10^3 = 1000)$

Number Patterns & Sequences 2

The nth Term of a Sequence

The nth term is a formula which enables us to generate any term within a particular sequence. Let our sequence of numbers have an **nth term = 2n + 2** where **n** is the position of the term, i.e. the first term has **n = 1**, the second term has **n = 2** and so on. This formula now enables us to generate any term simply by substituting our value for **n** into the formula.

$$\text{nth term} = 2n + 2$$
$$\text{1st term} = 2 \times 1 + 2 = 4$$
$$\text{2nd term} = 2 \times 2 + 2 = 6$$
$$\text{3rd term} = 2 \times 3 + 2 = 8$$
$$\text{4th term} = 2 \times 4 + 2 = 10$$
$$\text{100th term} = 2 \times 100 + 2 = 202$$

So our sequence of numbers would look like…

$$4, 6, 8, 10, \ldots$$

Finding the nth Term of an Arithmetic Sequence

A sequence where there is a common difference between the terms can be described by a linear algebraic expression. The general formula for the nth term of these sequences is…

$$\text{nth term} = an + b$$

where **a** is the common difference between the terms and **b** is an integer. The first thing you do is determine the value of **a**. To then find **b** substitute the value for the 1st term, **n = 1** and **a** into the formula.

Examples…

1 5,   7,   9,   11, …
    +2   +2   +2

These terms have a common difference of **+2** and so **a = 2**. If we take the 1st term then **n = 1** and it has a value of **5**. We then substitute these values into the formula…

$$\text{nth term} = an + b$$
$$5 = 2 \times 1 + b$$

… to give us **b = 5 − 2 = 3**

… Therefore **nth term = 2n + 3**

… To check **2nd term** = $2 \times 2 + 3 = 4 + 3 = 7$ ✓
3rd term = $2 \times 3 + 3 = 6 + 3 = 9$ ✓

2 20,   17,   14,   11, …
    -3   -3   -3

These terms have a common difference of **-3** and so **a = -3**. If we take the 1st term then **n = 1** and it has a value of **20**. We then substitute these values into the formula…

$$\text{nth term} = an + b$$
$$20 = -3 \times 1 + b$$

… to give us **b = 20 + 3 = 23**

… Therefore **nth term = -3n + 23**

… To check **2nd term** = $-3 \times 2 + 23 = -6 + 23 = 17$ ✓
3rd term = $-3 \times 3 + 23 = -9 + 23 = 14$ ✓

Plotting Points

All points are plotted on graph or squared paper. Usually your graph or squared paper is divided into four sections called **quadrants** by two lines known as the **x-axis**, which is a horizontal line, and the **y-axis**, which is a vertical line. The point where the two axes cross is called the **origin** (0,0).

The position of any plotted point is given by its **coordinates**. All coordinates are written as two numbers in a bracket separated by a comma, e.g. (3,4) (5,-2), where…

- The first number represents the **x** coordinate which is read going across horizontally to the right or left.
- The second number represents the **y** coordinate which is read vertically going up or down.

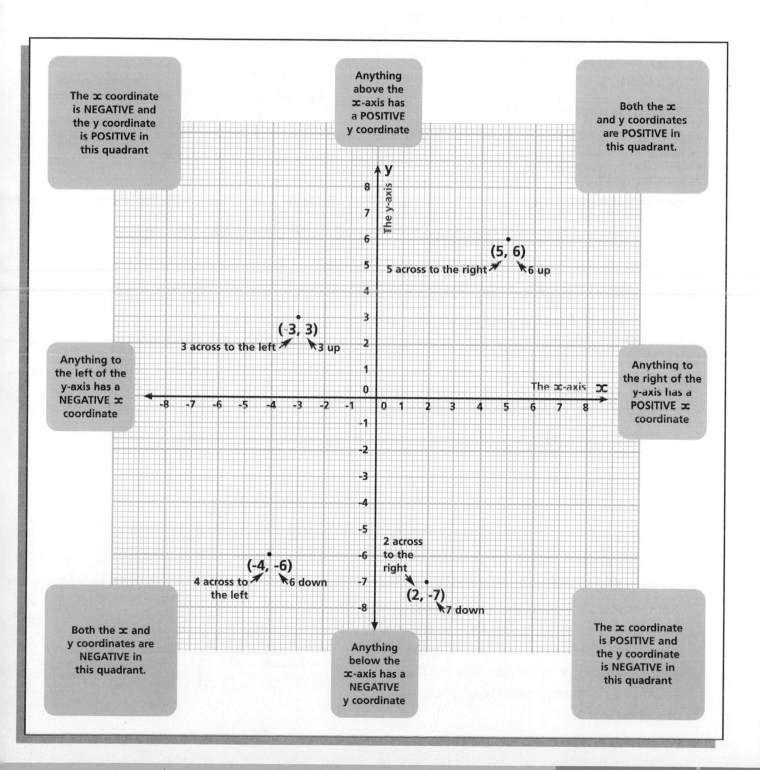

The **x** coordinate is NEGATIVE and the y coordinate is POSITIVE in this quadrant

Anything above the **x**-axis has a POSITIVE y coordinate

Both the **x** and y coordinates are POSITIVE in this quadrant.

(5, 6)
5 across to the right   6 up

(-3, 3)
3 across to the left   3 up

Anything to the left of the y-axis has a NEGATIVE **x** coordinate

Anything to the right of the y-axis has a POSITIVE **x** coordinate

The **x**-axis **x**

The y-axis

2 across to the right

(-4, -6)
4 across to the left   6 down

(2, -7)
7 down

Both the **x** and y coordinates are NEGATIVE in this quadrant.

Anything below the **x**-axis has a NEGATIVE y coordinate

The **x** coordinate is POSITIVE and the y coordinate is NEGATIVE in this quadrant

Graphs of Linear Functions 1

A linear function, e.g. $y = x$, $y = 2x - 1$, $y = 0.5x + 1$ (i.e. a function in which the highest power of x is x^1), will always give you a straight line graph when drawn.

To draw the graph of a linear function you only need to plot three points, for example...

---

1 Draw the graph of $y = 2x - 1$ for values of x between **-2** and **2** (this may be written $-2 \leqslant x \leqslant 2$).

Firstly, we need to pick 3 values of x, within the range, so that we can work out their **y** values. The two extreme values of x and one in the middle will do. Draw a table of results as follows:

Table of results for $y = 2x - 1$

x	-2	0	2
$2x$	(2 x -2 =) -4	(2 x 0 =) 0	(2 x 2 =) 4
-1	-1	-1	-1
$y = 2x - 1$	(-4 – 1 =) -5	(0 – 1 =) -1	(4 – 1 =) 3

We now have the coordinates of 3 points: (-2,-5), (0,-1) and (2,3) and can plot our graph.

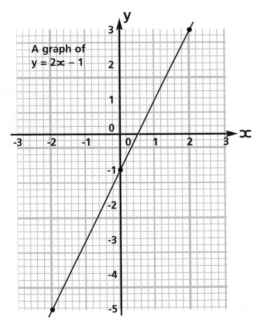

A graph of $y = 2x - 1$

Remember... Your straight line must pass through all 3 points! If it doesn't then one of your points is wrong. You can either check your table again or better still work out the coordinates of another point, e.g. when $x = -1$ or $x = 1$.

---

2 Draw the graph of $2y + x = 4$ for values of x between **-2** and **2**.

The first thing we have to do is rearrange the function to make **y** the subject. When we have done that we can draw a table of results.

$$2y + x = 4$$
$$2y + x - x = -x + 4 \quad \text{Subtract } x \text{ from both sides}$$
$$\frac{2y}{2} = \frac{-x}{2} + \frac{4}{2} \quad \text{Divide both sides by 2}$$

To give us ... $y = -0.5x + 2$

Table of results for $y = -0.5x + 2$

x	-2	0	2
$-0.5x$	(-0.5 x -2 =) 1	(-0.5 x 0 =) 0	(-0.5 x 2 =) -1
+ 2	+ 2	+ 2	+ 2
$y = -0.5x + 2$	(1 + 2 =) 3	(0 + 2 =) 2	(-1 + 2 =) 1

We now have the coordinates of 3 points: (-2,3), (0,2) and (2,1) and can plot our graph.

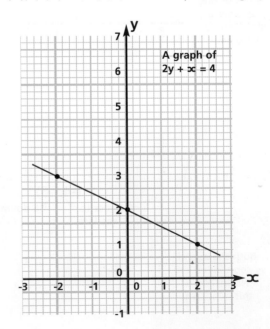

A graph of $2y + x = 4$

Graphs of Linear Functions 2

y = mx + c

The general equation for any straight line graph is **y = mx + c** where…

- … **m** is the value of the gradient. The gradient or slope of a line is a measure of the steepness of the line; the bigger the gradient the steeper the line. The gradient of a line can either be positive or negative, depending on which way the line slopes.
- … **c** is the value of the intercept. The intercept of a line is simply the **y** value at the point where the line crosses the y-axis.

Each of the graphs below have the same intercept (**c = +1**) but different gradients. The gradient of the lines in graph ❶ and ❷ are both positive with graph ❷ having a steeper slope (**m = 3** as compared to **m = 2** for graph ❶). The line in graph ❸ has a negative gradient (**m = -2**) and so it slopes in an opposite direction to the other two graphs.

Remember, a positive (+) gradient goes **up** from left to right (as we read) whereas a negative (-) gradient goes **down** from left to right.

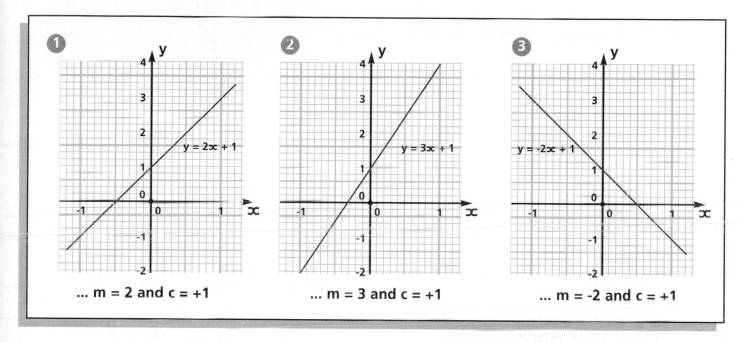

| ❶ … m = 2 and c = +1 | ❷ … m = 3 and c = +1 | ❸ … m = -2 and c = +1 |

Gradients of Parallel and Perpendicular Lines

Lines represented by equations that have the same gradient will be parallel even if each equation has a different intercept. For example, lines represented by **y = 2x + 2** and **y = 2x − 1** are parallel (see graph). Two lines are perpendicular, i.e. cross at 90°, if the product of their gradients is equal to -1. The line represented by $y = -\frac{1}{2}x + 3$ is perpendicular to both lines **y = 2x + 2** and **y = 2x − 1** since $-\frac{1}{2} \times 2 = -1$.

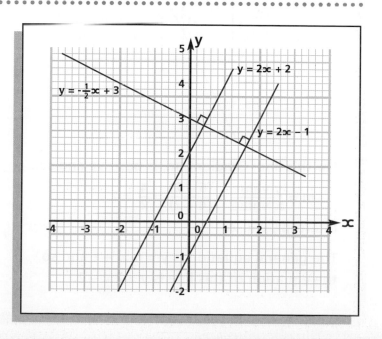

Finding the Equation of a Straight Line

In order to find the equation of a straight line graph all you have to do is find the gradient, **m**, and the intercept, **c**. Once you've got values for **m** and **c** you can substitute them into the general equation $y = mx + c$ to give the equation of the line. To find the gradient of a line pick two suitable points on your line and then complete a right-angled triangle as shown in the examples below. The gradient is given by the formula:

$$\text{GRADIENT} = \frac{y \text{ value}}{x \text{ value}}$$
For a positive gradient (see example 1)

$$\text{GRADIENT} = -\frac{y \text{ value}}{x \text{ value}}$$
For a negative gradient (see example 2)

Remember your measurement for the **y** value and the **x** value must be taken using the scales on the axes. You cannot measure them with a ruler.

Examples...

 ❶
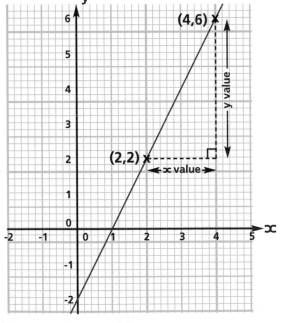

 ❷
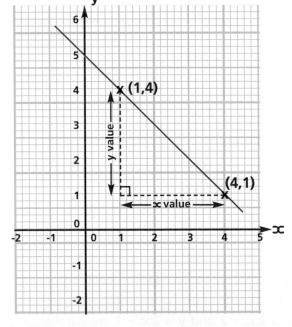

For this line: Gradient, m, is positive and so...

$$m = \frac{y \text{ value}}{x \text{ value}} = \frac{(6-2)}{(4-2)} = \frac{4}{2} = 2$$

Intercept, c = -2.

The general equation is $y = mx + c$ and so the equation of this line is $y = 2x - 2$.

For this line: Gradient, m, is negative and so...

$$m = -\frac{y \text{ value}}{x \text{ value}} = -\frac{(4-1)}{(4-1)} = -\frac{3}{3} = -1$$

Intercept, c = +5.

The general equation is $y = mx + c$ and so the equation of this line is $y = -1x + 5$ or $y = -x + 5$.

Three Special Graphs

❶ Graph of x = 'A NUMBER'

Examples are $x = 4, x = -3, x = 0$. The graphs of these equations are all VERTICAL LINES, i.e. they all go straight up and down.

For the graph x = 'A NUMBER' the x coordinates of all points on the line are always the same and equal to the 'number'. However, all the points will have different y coordinates (see drawn graphs).

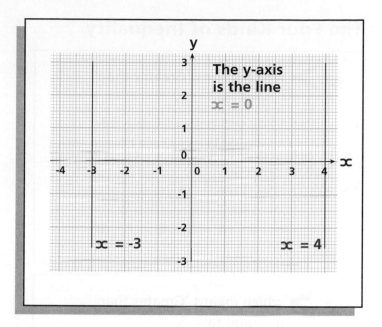

❷ Graph of y = 'A NUMBER'

Examples are $y = 3, y = -2, y = 0$. The graphs of these equations are all HORIZONTAL LINES, i.e. they all go straight across.

For the graph y = 'A NUMBER' the y coordinates of all points on the line are always the same and equal to the 'number'. This time, however, all the points will have different x coordinates (see drawn graphs).

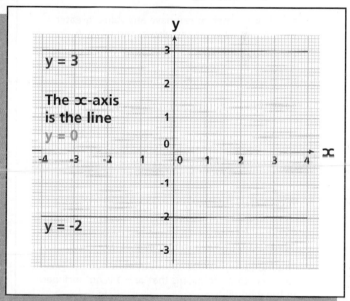

❸ Graphs of y = x and y = -x

The graphs of these equations are both DIAGONAL LINES but in opposite directions. Both lines always pass through the origin (0,0) and have a gradient of 1 and -1 respectively.

For the graph $y = x$, the x and y coordinates of a particular point on the line will be the same numerically and of the same sign (both + or both -).

For the graph $y = -x$, the x and y coordinates of a particular point on the line will be the same numerically, but of opposite signs (one + and one -).

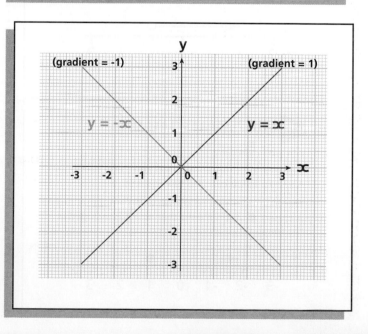

Linear Inequalities 1

The Four Kinds of Inequality

1 ❯ which means 'Greater than'

e.g. if $x > 4$ then x can have any value 'greater than' 4 but it can't have a value equal to 4.

This inequality can be shown using a number line. An open ○ circle means that $x = 4$ is not included in the inequality.

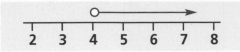

2 ❯❯ which means 'Greater than or Equal to'

e.g. if $x \geqslant 4$ then x can have any value 'greater than' 4 and its value can also be 'equal to' 4.

This inequality can be shown using a number line. A closed ● circle means that $x = 4$ is included in the inequality.

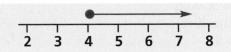

3 ❮ which means 'Less than'

e.g. if $x < 1$ then x can have any value 'less than' 1 but it can't have a value equal to 1.

This inequality can be shown using a number line. An open ○ circle means that $x = 1$ is not included in the inequality.

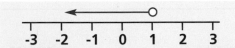

4 ❮❮ which means 'Less than or Equal to'

e.g. if $x \leqslant 1$ then x can have any value 'less than' 1 and its value can also be 'equal to' 1

This inequality can be shown using a number line. A closed ● circle means that $x = 1$ is included in the inequality.

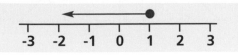

Number lines can also be used to show a combination of inequalities, for example...

1 $-2 \leqslant x < 5$

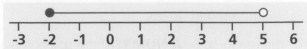

2 $1 < x \leqslant 4$

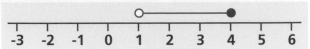

Solving Linear Inequalities

Solving these is just like solving linear equations except we have an inequality sign instead of the equal sign, for example...

1 $x + 2 < 8$
$x + 2 - 2 < 8 - 2$
$x < 6$

2 $3x - 3 \geqslant x + 7$
$3x - x - 3 \geqslant x - x + 7$
$2x - 3 + 3 \geqslant 7 + 3$
$\dfrac{2x}{2} \geqslant \dfrac{10}{2}$
$x \geqslant 5$

However, if you multiply or divide an inequality by a negative number then you must always reverse the direction of the inequality sign, for example...

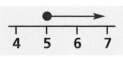

1 $-2x > 6$
$\dfrac{-2x}{-2} < \dfrac{6}{-2}$
$x < -3$

> Divide both sides by -2... inequality sign reverses direction (i.e. > becomes <)

2 $\dfrac{-x}{4} \leqslant 1.5$
$\dfrac{-x}{4} \times -4 \geqslant 1.5 \times -4$
$x \geqslant -6$

> Multiply both sides by -4... inequality sign reverses direction (i.e. ⩽ becomes ⩾)

Illustrating Linear Inequalities Graphically

Any linear inequality can be illustrated graphically.
All you have to do is…

1 Treat the inequality as an equation with an equal
(=) sign and draw its graph where a > or <
inequality is drawn as a dotted line and a
≥ or ≤ inequality is drawn as a solid line.

2 Label and shade in the region which satisfies the
inequality or inequalities.

Examples…

**1 Illustrate $x < 4$ graphically. Label and
shade the region which satisfies
the inequality.**

- Firstly draw the graph of $x = 4$ (see page 55),
remember to draw a dotted line.

- To determine the region which satisfies
the inequality, pick two points (we've labelled
them A and B) one on each side of the line
$x = 4$ (see graph).

- Label and shade the region which satisfies
the inequality.

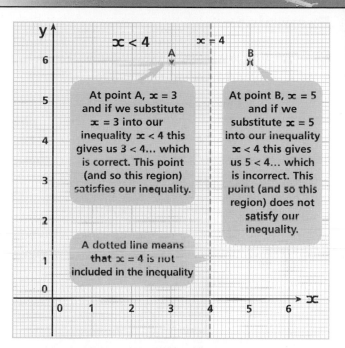

At point A, $x = 3$ and if we substitute $x = 3$ into our inequality $x < 4$ this gives us $3 < 4$… which is correct. This point (and so this region) satisfies our inequality.

At point B, $x = 5$ and if we substitute $x = 5$ into our inequality $x < 4$ this gives us $5 < 4$… which is incorrect. This point (and so this region) does not satisfy our inequality.

A dotted line means that $x = 4$ is not included in the inequality

**2 Illustrate $y ≥ 1$, $y < 2x$ and $x + y ≤ 6$
graphically. Label and shade the single region
that is satisfied by all of these inequalities.**

- Draw the graph of…

… **y = 1** (see page 55)

… **y = 2x**. Make a simple table of results.

x	0	2	4
$y = 2x$	0	4	8

… **x + y = 6**. Rearrange to give **y** on its own
and make a simple table of results.

x	0	3	6
$y = 6 - x$	6	3	0

- Label and shade the region which satisfies
all three inequalities.

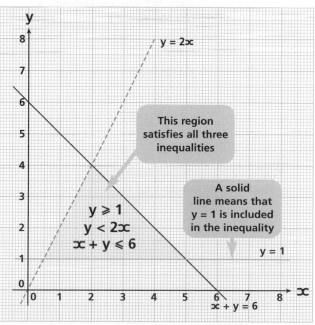

This region satisfies all three inequalities

A solid line means that $y = 1$ is included in the inequality

$y ≥ 1$
$y < 2x$
$x + y ≤ 6$

Simultaneous Equations 1

Solving Simultaneous Equations by Elimination

Simultaneous equations are where you work with two related equations (e.g. where x and y represent the same thing in both equations) at the same time. Individually the equations do not provide enough information for you to solve them (i.e. to find the values of x and y). However when you work with both of them 'simultaneously' they provide enough combined information to solve them.

To solve simultaneous equations we need both equations to have the same number in front of EITHER of the two unknown letters (e.g. the x's or the y's).

- If these two NUMBERS have the SAME SIGN (i.e. both positive or both negative) then we SUBTRACT ONE EQUATION FROM THE OTHER.
- If these two NUMBERS have DIFFERENT SIGNS (i.e. one is positive and one is negative) then we ADD THE TWO EQUATIONS TOGETHER.

Examples...

1 Solve the two simultaneous equations:
$4x + 2y = 14$ and $x + 2y = 8$.

> Label the equations **1** and **2**

$$4x + 2y = 14 \quad \textbf{1}$$
$$x + 2y = 8 \quad \textbf{2}$$

> Equations **1** and **2** both have the same number of y's with the same sign (+), so we subtract one equation from the other to remove the y's.

Subtract **1** – **2**
$$4x + 2y = 14 \quad \textbf{1}$$
$$- (x + 2y = 8) \quad \textbf{2}$$
$$\overline{3x \qquad = 6}$$

To give us: $x = 2$

> To find the value of y, SUBSTITUTE $x = 2$ back into equation **1** or **2** (pick the easiest). If we take equation **2** ...

$$x + 2y = 8$$
$$2 + 2y = 8$$
$$2y = 8 - 2$$
$$2y = 6$$
$$y = 3$$

> To check our solutions substitute $x = 2$ and $y = 3$ into equation **1** since we used equation **2** to find the value of y.
>
> $4x + 2y = 14$
> $4 \times 2 + 2 \times 3 = 14$
> $8 + 6 = 14$ and that's correct. ✔

2 Solve the two simultaneous equations:
$2p + 3q = 13$ and $3p - q = 3$

> Label the equations **1** and **2**

$$2p + 3q = 13 \quad \textbf{1}$$
$$3p - q = 3 \quad \textbf{2}$$

> Multiply equation **2** by 3 so that this equation now has the same number of q's as equation **1**. Equation **2** now becomes equation **3**.

2 x 3 ⟶ $9p - 3q = 9 \quad \textbf{3}$

> Equations **1** and **3** now have the same number of q's but with different signs so we add the two equations together to remove the q's.

Add **1** + **3**
$$2p + 3q = 13 \quad \textbf{1}$$
$$+ (9p - 3q = 9) \quad \textbf{3}$$
$$\overline{11p \qquad = 22}$$

To give us: $p = 2$

> To find the value of q, SUBSTITUTE $p = 2$ into equation **1**, **2** or **3**. If we take the equation **1** ...

$$2p + 3q = 13$$
$$2 \times 2 + 3q = 13$$
$$3q = 13 - 4$$
$$3q = 9$$
$$q = 3$$

> To check our solutions substitute $p = 2$ and $q = 3$ into equation **2** or **3** since we used equation **1** to find the value of q. If we take equation **2** ...
>
> $3p - q = 3$
> $3 \times 2 - 3 = 3$
> $6 - 3 = 3$ and that's correct. ✔

Simultaneous Equations 2

Solving Simultaneous Equations Graphically

Simultaneous equations can also be solved graphically. Both equations when plotted give us straight line graphs which will cross each other.

At the point of intersection the corresponding x and y values are the solutions of the simultaneous equations.

Examples...

1 Solve graphically the two simultaneous equations $x + y = 5$ and $2x + y = 7$ where the value of x lies in the range $1 \leqslant x \leqslant 5$ (i.e. somewhere between **1** and **5**).

> For each equation we need 3 corresponding values of x and y which can then be plotted so that we can draw our graph. This may mean rearranging an equation to give us y on its own.

OUR FIRST EQUATION

$x + y = 5$	x	1	3	5
$\therefore y = 5 - x$	$y = 5 - x$	(5–1=) 4	(5–3=) 2	(5–5=) 0

OUR SECOND EQUATION

$2x + y = 7$	x	1	3	5
$\therefore y = 7 - 2x$	$y = 7 - 2x$	(7–2x1=) 5	(7–2x3=) 1	(7–2x5=) -3

> To check our solutions substitute $x = 2$ and $y = 3$ into our two equations: $x + y = 5$: $2 + 3 = 5$ ✓
> and $2x + y = 7$: $2 \times 2 + 3 = 7$: $4 + 3 = 7$ ✓

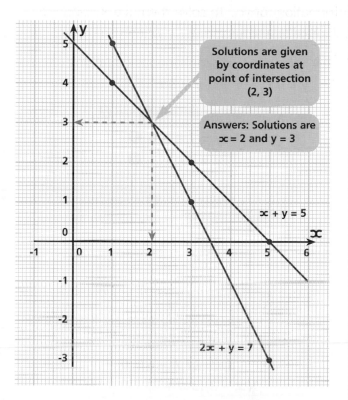

> Solutions are given by coordinates at point of intersection (2, 3)

> Answers: Solutions are $x = 2$ and $y = 3$

2 Solve graphically the two simultaneous equations $4x = y + 3$ and $2x + 2y = 9$ where the value of x lies in the range $0 \leqslant x \leqslant 2$ (i.e. somewhere between **0** and **2**).

OUR FIRST EQUATION

$4x = y + 3$	x	0	1	2
$y = 4x - 3$	$y = 4x - 3$	(4x0–3=)-3	(4x1–3=) 1	(4x2–3=) 5

OUR SECOND EQUATION

$2x + 2y = 9$	x	0	1	2
$2y = 9 - 2x$		$\frac{(9-2x0)}{2}=$	$\frac{(9-2x1)}{2}=$	$\frac{(9-2x2)}{2}=$
$y = \dfrac{9 - 2x}{2}$	$y = \frac{9-2x}{2}$	4.5	3.5	2.5

> To check our solutions substitute $x = 1.5$ and $y = 3$ into our two equations:
> $4x = y + 3$: $4 \times 1.5 = 3 + 3$: $6 = 3 + 3$ ✓
> and $2x + 2y = 9$: $2 \times 1.5 + 2 \times 3 = 9$: $3 + 6 = 9$ ✓

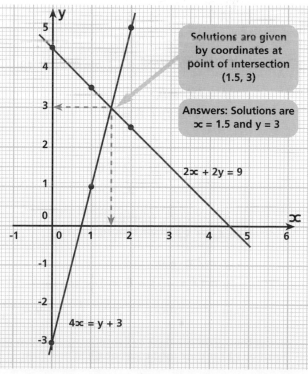

> Solutions are given by coordinates at point of intersection (1.5, 3)

> Answers: Solutions are $x = 1.5$ and $y = 3$

Simultaneous Equations 3

Solving Simultaneous Linear and Quadratic Equations by Elimination

You are also expected to be able to solve by elimination two simultaneous equations with two unknowns (i.e. x and y), where one equation is linear in each unknown (e.g. $y = 7x - 6$, $2x + y = 11$) and the other equation is linear in one unknown and quadratic in the other (e.g. $y = 3x^2$), or is of the form $x^2 + y^2 = r^2$ (e.g. $x^2 + y^2 = 25$ which is the same as $x^2 + y^2 = 5^2$). $x^2 + y^2 = r^2$ is the equation of a circle of radius r, centred at the origin $(0,0)$ (see next page for graph).

Examples...

1 Solve the two simultaneous equations:

$$y = 7x + 6 \text{ and } y = 3x^2$$

Label the equations **1** and **2**.

$$y = 7x + 6 \text{ ❶}$$
$$y = 3x^2 \text{ ❷}$$

Since both equations have y on its own as the subject, substitute y from equation **2** into equation **1**. Doing this will remove y from equation **1**.

$$3x^2 = 7x + 6$$
$$3x^2 - 7x - 6 = 0$$

What we have now is a quadratic equation. Always solve by factorisation if possible (see page 45). However you can solve by completing the square or by using the quadratic formula (see pages 46 and 47).

$$3x^2 - 7x - 6 = 0$$
$$(x - 3)(3x + 2) = 0$$

And so... $x - 3 = 0$
$$x = 3$$

To find the value of y, substitute $x = 3$ into equation **1** ...

$$y = 7x + 6$$
$$y = (7 \times 3) + 6 = 27$$

One pair of solutions is $x = 3$ and $y = 27$.

or... $3x + 2 = 0$
$$x = -\tfrac{2}{3}$$

To find the value of y, substitute $x = -\tfrac{2}{3}$ into equation **1** or **2**. If we take equation **1** ...

$$y = 7x + 6$$
$$y = (7 \times -\tfrac{2}{3}) + 6 = 1\tfrac{1}{3}$$

The other pair of solutions is $x = -\tfrac{2}{3}$ and $y = 1\tfrac{1}{3}$.

Always check your solutions by substituting them into equation **2** since we used equation **1** to find them.

2 Solve the two simultaneous equations:

$$2x + y = 11 \text{ and } x^2 + y^2 = 25$$

Label the equations **1** and **2**.

$$2x + y = 11 \text{ ❶}$$
$$x^2 + y^2 = 25 \text{ ❷}$$

Rearrange equation **1** to make y the subject.

$$y = 11 - 2x$$

We can now substitute y from equation **1** into equation **2**. Doing this will remove y from equation **2**.

$$x^2 + (11 - 2x)^2 = 25$$
$$x^2 + (11 - 2x)(11 - 2x) = 25$$
$$x^2 + 121 - 22x - 22x + 4x^2 = 25$$
$$x^2 + 4x^2 - 22x - 22x + 121 - 25 = 0$$
$$5x^2 - 44x + 96 = 0$$
$$(x - 4)(5x - 24) = 0$$

And so... $x - 4 = 0$
$$x = 4$$

To get y substitute $x = 4$ into rearranged equation **1**.

$$y = 11 - 2x$$
$$y = 11 - (2 \times 4) = 3$$

One pair of solutions is $x = 4$ and $y = 3$.

or... $5x - 24 = 0$
$$5x = 24$$
$$x = \frac{24}{5} = 4.8$$

To get y substitute $x = 4.8$ into rearranged equation **1**.

$$y = 11 - 2x$$
$$y = 11 - (2 \times 4.8) = 1.4$$

The other pair of solutions is $x = 4.8$ and $y = 1.4$

Always check your solutions by substituting them into equation **2** since we used equation **1** to find them.

Simultaneous Equations 4

Solving Simultaneous Linear and Quadratic Equations Graphically

Approximate solutions to simultaneous linear and quadratic equations can also be found graphically.

When drawn there are two points of intersection (there are exceptions, see NOTE) to give us two pairs of solutions.

Examples...

1 Solve graphically the two simultaneous equations $y - x = 4$ and $y = 2x^2$ where the value of x lies in the range $-2 \leqslant x \leqslant 2$.

OUR FIRST EQUATION

$y - x = 4$

$\therefore y = x + 4$

x	-2	0	2
$y = x + 4$	2	4	6

OUR SECOND EQUATION

$y = 2x^2$

(see next page)

x	-2	-1	0	1	2
$y = 2x^2$	8	2	0	2	8

One pair of solutions is $x = -1.2$ and $y = 2.8$

The other pair of solutions is $x = 1.7$ and $y = 5.7$

> Remember to check your solutions by substituting the values of x and y into your two equations.

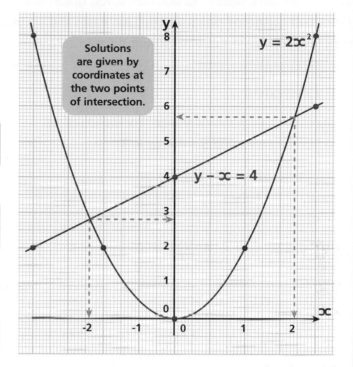

Solutions are given by coordinates at the two points of intersection.

$y = 2x^2$

$y - x = 4$

2 Solve graphically the two simultaneous equations $x + y = -1$ and $x^2 + y^2 = 9$ where the value of x lies in the range $-3 \leqslant x \leqslant 3$.

OUR FIRST EQUATION

$x + y = -1$

$\therefore y = -x - 1$

x	-3	0	3
$y = -x - 1$	2	-1	-4

OUR SECOND EQUATION

The equation is $x^2 + y^2 = 9$ or $x^2 + y^2 = 3^2$, i.e. the equation of a circle of radius 3 units centred at the origin (0,0).

One pair of solutions is $x = -2.6$ and $y = 1.6$

The other pair of solutions is $x = 1.6$ and $y = -2.6$

> Remember to check your solutions by substituting the values of x and y into your two equations.

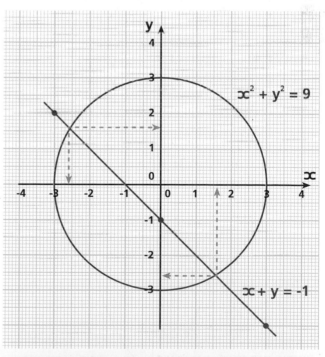

$x^2 + y^2 = 9$

$x + y = -1$

NOTE: It is possible to get only ONE pair of solutions. This would occur if the line was a tangent to the curve or circle.

Graphs of Quadratic Functions 1

Drawing Graphs of Quadratic Functions

A quadratic function is one in which the highest power is x^2.
Examples would include $y = x^2$, $y = x^2 - 5$, $y = x^2 - 2x$,
$y = x^2 + 4$. These functions always produce a curved graph
such as the ones below. To draw a curved graph we need to
plot a **full range of points** as this increases the accuracy of
our curve (compare this with straight line graphs).

Examples...

1 Draw the graph of $y = x^2$ for values of x
between **-3** and **3** ($-3 \leqslant x \leqslant 3$).

Table of results for $y = x^2$

x	-3	-2	-1	0	1	2	3
	$(-3)^2=$	$(-2)^2=$	$(-1)^2=$	$(0)^2=$	$(1)^2=$	$(2)^2=$	$(3)^2=$
$y = x^2$	9	4	1	0	1	4	9

We now have the coordinates of 7 points, so
we can draw our graph.

Remember... Your curve must be smooth
with no wobbly bits in it and it must pass
through all of the points plotted.

2 Draw the graph of $y = x^2 - 2x - 2$ for values
of x between **-2** and **4** ($-2 \leqslant x \leqslant 4$).

Table of results for $y = x^2 - 2x - 2$

x	-2	-1	0	1	2	3	4
x^2	$(-2)^2=$ 4	$(-1)^2=$ 1	$(0)^2=$ 0	$(1)^2=$ 1	$(2)^2=$ 4	$(3)^2=$ 9	$(4)^2=$ 16
$-2x$	-2x-2= +4	-2x-1= +2	-2x0= 0	-2x1= -2	-2x2= -4	-2x3= -6	-2x4= -8
-2	-2	-2	-2	-2	-2	-2	-2
$y=x^2-2x-2$	6	1	-2	-3	-2	1	6

Again we have the coordinates of 7 points, so
we can draw our graph.

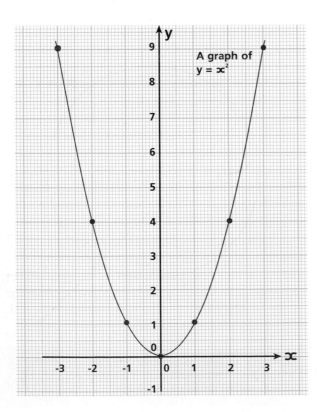

A graph of $y = x^2$

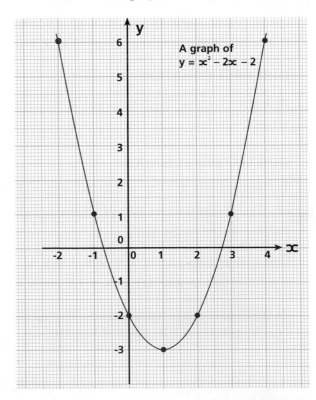

A graph of $y = x^2 - 2x - 2$

Graphs of Quadratic Functions 2

Solving Quadratic Equations Graphically

Approximate solutions to a quadratic equation can be found from the graph of the corresponding quadratic function, for example...

The graph of $y = x^2 - 2x - 2$ (see previous page) can be used to solve the quadratic equation $x^2 - 2x - 2 = 0$.

Our quadratic equation is... $\boxed{x^2 - 2x - 2} = 0$

Our graph is .... $y = \boxed{x^2 - 2x - 2}$

Since our quadratic equation corresponds to our quadratic function (i.e. $x^2 - 2x - 2$ is common to both) then approximate solutions are simply the x values where the graph crosses the line $y = 0$, i.e. the x-axis. The two approximate solutions to the equation $x^2 - 2x - 2 = 0$ are $x = -0.7$ and $x = 2.7$ (see graph).

A graph of a quadratic function can also be used to solve a wide variety of quadratic equations. However you will probably have to rearrange the quadratic equation so that it corresponds to the quadratic function before any solutions can be obtained, for example ...

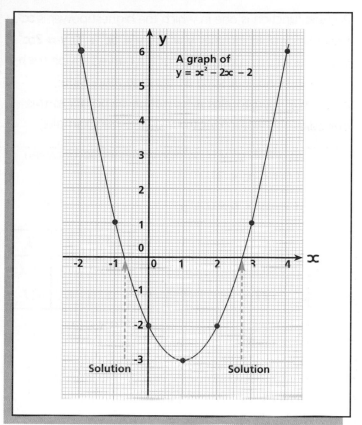

A graph of $y = x^2 - 2x - 2$

Solution   Solution

---

1 Use the graph of $y = x^2 - 2x - 2$ to solve the quadratic equation $x^2 - 2x - 5 = 0$.

Firstly, we need to rearrange the quadratic equation so that it corresponds to the quadratic function.

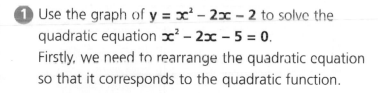

Our equation is...	$x^2 - 2x - 5 = 0$
Add 3 to both sides	$x^2 - 2x - 5 + 3 = 0 + 3$
Our equation now becomes...	$\boxed{x^2 - 2x - 2} = 3$

Our graph is .... $y = \boxed{x^2 - 2x - 2}$

The solutions are the x values where the graph crosses the line $y = 3$: $x = -1.4$ and $x = 3.4$ (see graph)

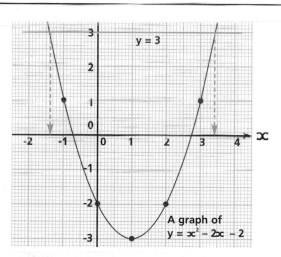

$y = 3$

A graph of $y = x^2 - 2x - 2$

2 Use the graph of $y = x^2 - 2x - 2$ to solve the quadratic equation $x^2 - 2x - 1 = 0$.

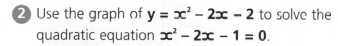

Our equation is...	$x^2 - 2x - 1 = 0$
Subtract 1 from both sides	$x^2 - 2x - 1 - 1 = 0 - 1$
Our equation now becomes...	$\boxed{x^2 - 2x - 2} = -1$

Our graph is .... $y = \boxed{x^2 - 2x - 2}$

This time the solutions are the x values where the graph crosses the line $y = -1$: $x = -0.4$ and $x = 2.4$ (see graph)

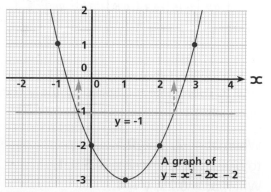

$y = -1$

A graph of $y = x^2 - 2x - 2$

Graphs of Other Functions 1

Graphs of Simple Cubic Functions

A cubic function is one in which the highest power is x^3.
Examples would include $y = x^3$, $y = x^3 + 1$, $y = 2x^3$.
These functions always produce a curved graph with
a double bend in it such as the one alongside.
Again we need to plot a full range of points as this
increases the accuracy of our curve, for example…

Draw the graph of $y = x^3$ for values of x between
-3 and 3 ($-3 \leqslant x \leqslant 3$).

Table of results for $y = x^3$

x	-3	-2	-1	0	1	2	3
	$(-3)^3 =$	$(-2)^3 =$	$(-1)^3 =$	$(0)^3 =$	$(1)^3 =$	$(2)^3 =$	$(3)^3 =$
$y = x^3$	-27	-8	-1	0	1	8	27

Since we now have the coordinates of 7 points we
can draw our graph. As before, your curve must be
smooth with no wobbly bits in it and it must pass
through all of the points plotted.

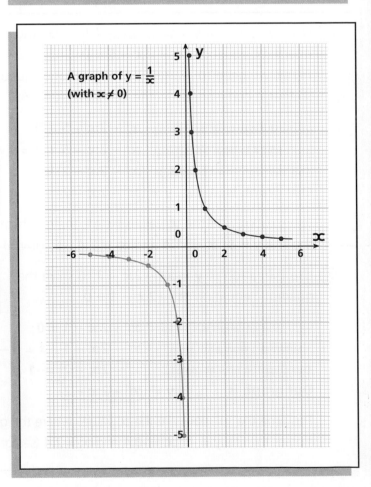

Graph of the Reciprocal Function $y = \frac{1}{x}$ with $x \neq 0$

The reciprocal function $y = \frac{1}{x}$ with $x \neq 0$ gives us
two separate curves.

For negative values of x…

x	-5	-4	-3	-2	-1	-0.5	-0.$\dot{3}$	-0.25	-0.2
$y = \frac{1}{x}$	-0.2	-0.25	-0.$\dot{3}$	-0.5	-1	-2	-3	-4	-5

For positive values of x…

x	0.2	0.25	0.$\dot{3}$	0.5	1	2	3	4	5
$y = \frac{1}{x}$	5	4	3	2	1	0.5	0.$\dot{3}$	0.25	0.2

With the function $y = \frac{1}{x}$, for any point on either
curve the x coordinate multiplied by the y coordinate
always equals 1.

e.g. When $x = -2$, $y = -0.5$…
to give us $x \times y = -2 \times -0.5 = 1$
When $x = 0.25$, $y = 4$…
to give us $x \times y = 0.25 \times 4 = 1$

Graphs of Other Functions 2

Graphs of the Exponential Function $y = k^x$

You need to be able to draw the graph of the exponential function $y = k^x$ for integer values of x and simple positive values of k. For example...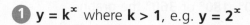

1 $y = k^x$ where $k > 1$, e.g. $y = 2^x$

x	0	1	2	3	4
$y = 2^x$	$(2)^0 =$ 1	$(2)^1 =$ 2	$(2)^2 =$ 4	$(2)^3 =$ 8	$(2)^4 =$ 16

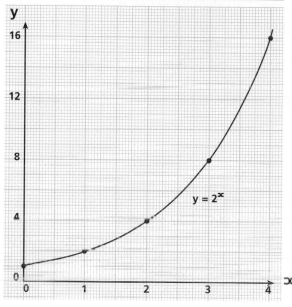

2 $y = k^x$ where $k < 1$, e.g. $y = (0.5)^x$

x	0	1	2	3	4
$y = (0.5)^x$	$(0.5)^0 =$ 1	$(0.5)^1 =$ 0.5	$(0.5)^2 =$ 0.25	$(0.5)^3 =$ 0.125	$(0.5)^4 =$ 0.0625

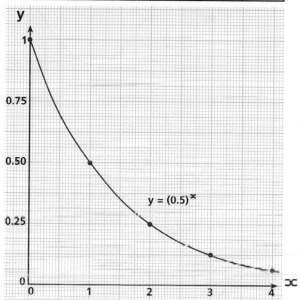

Example...

The graph below is known to fit the relationship $y = ab^x$. Use the graph to find the value of **a** and **b** and the relationship.

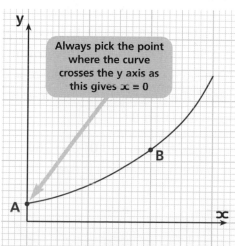

Always pick the point where the curve crosses the y axis as this gives $x = 0$

This time we need to take values for x and y from our graph and then substitute them into the relationship, $y = ab^x$.

At point A:

$x = 0$ (curve crosses the y-axis) and $y = 3$

$y = ab^x$ gives us: $3 = ab^0$

$3 = a \times 1$ — Anything to the power of 0 = 1

and so... **a = 3**

At point B:

$x = 2$ and $y = 12$ (we know that $a = 3$)

$y = ab^x$ gives us: $12 = 3b^2$

$4 = b^2$

and so... $b = \sqrt{4} = 2$

The relationship is: $y = 3 \times 2^x$

Graphs of Other Functions 3

Graph of y = sin x

If we plot the graph of **y = sin x** for values of **x** from **0°** to **360°** we get the following:

x (°)	0	30	60	90	120	150	180	210	240	270	300	330	360
y = sin x	0	0.5	0.87	1	0.87	0.5	0	-0.5	-0.87	-1	-0.87	-0.5	0

y = sin x

The graph is wave-like, where the maximum value of **sin x** is **1** when **x = 90°** and the minimum value of **sin x** is **-1** when **x = 270°**. When **x = 0°**, **180°** or **360°** the graph crosses the x-axis and **sin x** is equal to **0**.

The graph opposite is for values of **x** from **0°** to **360°**. It is also possible to extend beyond this range, i.e. values of **x** less than **0°** and greater than **360°**. The pattern or cycle repeats itself every 360°.

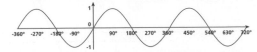

Graph of y = cos x

If we plot the graph of **y = cos x** for values of **x** from **0°** to **360°** we get the following:

x (°)	0	30	60	90	120	150	180	210	240	270	300	330	360
y = cos x	1	0.87	0.5	0	-0.5	-0.87	-1	-0.87	-0.5	0	0.5	0.87	1

y = cos x

The graph is wave-like (same as the graph of **y = sin x**, except that its been shifted 90° to the left), where the maximum value of **cos x** is **1** when **x = 0°** or **360°** and the minimum value of **cos x** is **-1** when **x = 180°**. When **x = 90°** or **270°** the graph crosses the x-axis and **cos x** is equal to **0**.

Also, as for the graph of **y = sin x**, we can extend beyond **0°** and **360°**. The pattern again repeats itself every 360°.

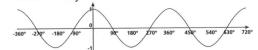

Graph of y = tan x

If we plot the graph of **y = tan x** for values of **x** from **0°** to **360°** we get the following:

x (°)	0	30	60	90	120	150	180	210	240	270	300	330	360
y = tan x	0	0.58	1.73	INF	-1.73	-0.58	0	0.58	1.73	INF	-1.73	-0.58	0

y = tan x

The graph is not wave-like like the above graphs. When **x = 90°** or **270°** then **tan x** is **infinity (INF)** which is a never-ending value. When **x = 0°**, **180°** or **360°** the graph crosses the x-axis and **tan x** is equal to **0**.

Also, we can extend beyond **0°** and **360°**. Once again the pattern repeats itself every 360°.

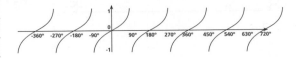

Transformations of Functions 1

If **y = an expression involving x** then we can say that **y** is a function of **x** or **y = f(x)**. You may well be asked to draw the graph of **another function** which is related to the graph of a function which has been already drawn for you, for example…

---

① **y = af(x)**… graph is **pulled/pushed** in the **y direction**

If **a > 1**, e.g. **y = 2x²**, then all the points are pulled upwards in the **y** direction by a factor of **a** (opposite **a = 2**)

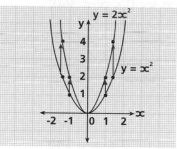

If **a < 1**, e.g. **y = 0.5x²**, then all the points are pushed downwards in the **y** direction by a factor of **a** (opposite **a = 0.5**)

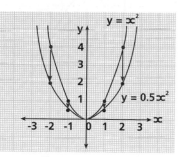

---

② **y = f(ax)**… graph is **pushed/pulled** in the **x direction**

If **a > 1**, e.g. **y = (2x)²**, then all the points are pushed inwards in the **x** direction by a factor of $\frac{1}{a}$ (opposite $\frac{1}{a} = \frac{1}{2} = 0.5$)

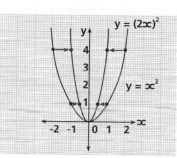

If **a < 1**, e.g. **y = (0.5x)²**, then all the points are pulled outwards in the **x** direction by a factor of $\frac{1}{a}$ (below $\frac{1}{a} = \frac{1}{0.5} = 2$)

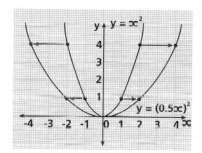

---

③ **y = f(x + a)**… graph is **translated** in the **x direction**

If **a is positive**, e.g. **y = (x + 1)²**, then all the points are translated to the left by the value of **a** (opposite **a = 1**)

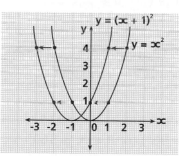

If **a is negative**, e.g. **y = (x − 1)²**, then all the points are translated to the right by the value of **a** (opposite **a = -1**)

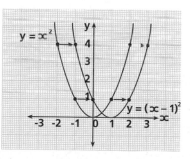

---

④ **y = f(x) + a**… graph is **translated** in the **y direction**

If **a is positive**, e.g. **y = x² + 1**, then all the points are translated upwards by the value of **a** (opposite **a = 1**)

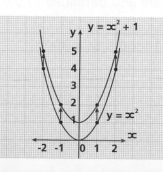

If **a is negative**, e.g. **y = x² − 1**, then all the points are translated downwards by the value of **a** (opposite **a = -1**)

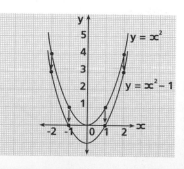

Transformations of Functions 2

The examples below all refer to the transformation of the **sine** and **cosine** functions. Each example includes either the graph $y = \sin x$ or $y = \cos x$ so that you can make comparisons.

①

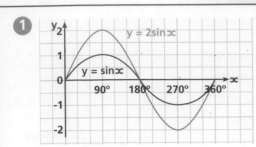

$y = a\sin x$

e.g. $y = 2\sin x$ for $0 \leqslant x \leqslant 360°$

Since **a > 1** then all the points are pulled outwards in the **y** direction by a factor of **2**. Note that if **a < 1** (e.g. $y = 0.5\sin x$) then all the points are pushed inwards in the **y** direction.

②

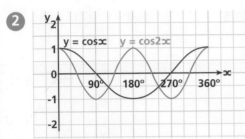

$y = \cos a x$

e.g. $y = \cos 2x$ for $0 \leqslant x \leqslant 360°$

Since **a > 1** then all the points are pulled inwards in the **x** direction by a factor of **2**. Note that if **a < 1** (e.g. $y = \cos 0.5x$) then all the points are stretched outwards in the **x** direction.

③

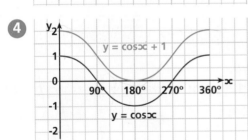

$y = \sin(x + a)$

e.g. $y = \sin(x + 90°)$ for $0 \leqslant x \leqslant 360°$

Since **a** is positive then all the points are translated to the left by the value of **a**, i.e. **90°**. Note that if **a** is negative (e.g. $y = \sin(x - 90°)$) then all the points are translated to the right.

④

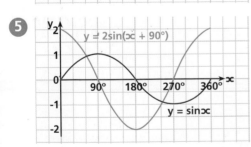

$y = \cos x + a$

e.g. $y = \cos x + 1$ for $0 \leqslant x \leqslant 360°$

Since **a** is positive then all the points are translated upwards by the value of **a**, i.e. **1**. Note that if **a** is negative (e.g. $y = \cos x - 1$) then all the points are translated downwards.

⑤

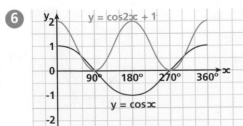

These final two examples show a combination of transformations.
$y = 2\sin(x + 90°)$ for $0 \leqslant x \leqslant 360°$

This is a combination of the transformations in examples **①** and **③** above, i.e. all the points are pulled outwards in the **y** direction **and** are also translated to the left!

⑥

$y = \cos 2x + 1$ for $0 \leqslant x \leqslant 360°$

This is a combination of the transformations in examples **②** and **④** above, i.e. all the points are pulled inwards in the **x** direction **and** are also translated upwards!

Distance - Time Graphs

These are also known as travel graphs, where distance is always plotted on the vertical axis and time is always plotted on the horizontal axis. For a distance-time graph the slope or gradient is always equal to the speed. If there is no slope, there is no movement, i.e. speed = 0.

Example...

A boy sets off from home riding his bike to go to a friend's house. A distance-time graph of his journey is shown opposite. Describe his movement between a) 0 and A, b) A and B, and c) B and C.

a) Between 0 and A the boy is cycling with constant speed given by the slope or gradient.

$$Speed = \frac{5km}{0.5h} = 10km/h$$

b) Between A and B there is no movement, i.e. the boy has stopped cycling for 1 hour.
Speed = 0 since there is no slope or gradient.

c) Between B and C the boy is again cycling with constant speed given by the slope or gradient.

$$Speed = \frac{2km}{0.5h} = 4km/h.$$

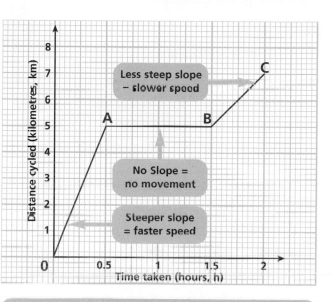

Notice that at the end of the journey he is 7km from home and it is 2 hours since he set off.

Velocity - Time Graphs

These are very similar to distance-time graphs except that the slope or gradient is equal to the acceleration. This time, if there is no slope then the object is moving with constant velocity (or speed) as can be seen on this graph. Between 0 and A the object is moving with a constant acceleration given by the uniform slope or gradient.

Acceleration = $\frac{10m/s}{20s}$ = 0.5m/s^2

(m/s^2 is metres per second squared or metres per second per second, the unit for acceleration)

Between A and B the object is moving with a constant velocity (or speed) of 10m/s, i.e. it is not accelerating.

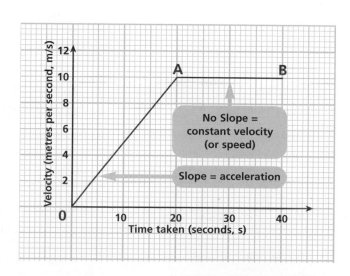

Other Graphs 2

Conversion Graphs

These are used to convert values of one quantity into another, e.g. pounds sterling (£) into euros (€) or any other currency, miles into kilometres and so on.

Example

Draw a conversion graph for pounds sterling and euros up to £30 if £1 = €1.60. From your graph convert…

a) £17 into euros

b) €40 into pounds sterling

Before we can draw our graph we need a table of values for pounds sterling and euros.

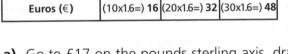

Pounds sterling (£)	10	20	30
Euros (€)	(10x1.6=) **16**	(20x1.6=) **32**	(30x1.6=) **48**

a) Go to £17 on the pounds sterling axis, draw a dotted line across (➤) to the graph and then down (▼) to the euros axis. **£17 = €27.**

b) Go to €40 on the euros axis, draw a dotted line up (▲) to the graph and then across (◄) to the pounds sterling axis. **€40 = £25.**

Graphs that Describe Real Life Situations

Here are 6 examples (there are numerous others) where a graph can be used to show a real life situation.

All of these graphs require common sense in their interpretation as each one is very different.

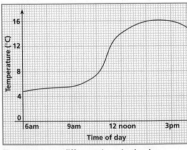

Temperature at different times in the day

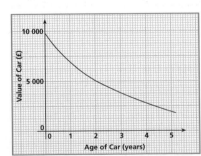

Value of a car as it gets older

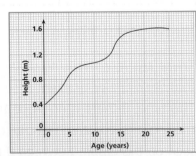

Height of a girl as she gets older

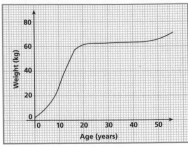

Weight of a man as he gets older

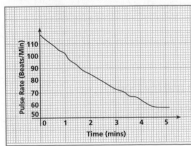

Pulse rate of a runner after the race is over

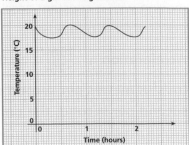

Temperature of a thermostatically controlled room over a period of time

Direct and Inverse Proportion

Direct Proportion

Two quantities are in direct proportion if when we double (or treble) one of the quantities then the other quantity also doubles (or trebles). If two quantities **y** and **x** are in direct proportion then a graph of values of **y** against values of **x** would be a straight line.

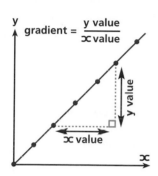

gradient = $\frac{y \text{ value}}{x \text{ value}}$

We can say that...

y is directly proportional to **x** or **y∝x** which means that **y = kx** where **k** is a constant, i.e. a number that doesn't change and is given by the **gradient** of the graph.

It is also possible for **y∝x²**, **y∝x³** and so on.

Example...

The distance, **d**, travelled by an accelerating car which was initially stationary is directly proportional to the **square** of the time, **t**, of travel. The table below gives values of **d** and **t**. What is the relationship between **d** and **t**?

d(m)	0	3	12	27
t(s)	0	1	2	3
t²(s²)	(0²=) 0	(1²=) 1	(2²=) 4	(3²=) 9

Since **d∝t²** or **d = kt²** we've added one row, **t²**, to our table (shown in red). A graph of **d** against **t²** will give us a straight line.

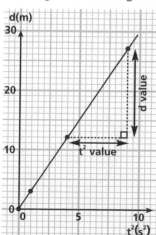

gradient = k

$= \frac{d \text{ value}}{t^2 \text{ value}}$

$= \frac{(27 - 12)}{(9 - 4)}$

$= \frac{15}{5}$

$= 3$

Relationship is:

d = 3t²

Inverse Proportion

Two quantities are in inverse proportion if when we double (or treble) one of the quantities then the other quantity halves (or becomes a third).
If two quantities **y** and **x** are in inverse proportion then a graph of values of **y** against values of $\frac{1}{x}$ would be a straight line.

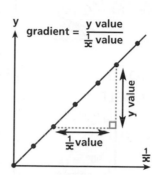

gradient = $\frac{y \text{ value}}{\frac{1}{x} \text{ value}}$

We can say that...

y is inversely proportional to **x** or **y∝$\frac{1}{x}$** which means that **y = $\frac{k}{x}$** where **k** is a constant and is given by the **gradient** of the graph.

It is also possible for **y∝$\frac{1}{x^2}$**, **y∝$\frac{1}{x^3}$** and so on.

Example...

The density, **d**, of objects of fixed mass is inversely proportional to their volume, **V**. The table below gives values of **d** and **V**. What is the relationship between **d** and **V**?

d(kg/m³)	1	2	5	10
V(m³)	10	5	2	1
$\frac{1}{V}$ (m⁻³)	($\frac{1}{10}$ =) 0.1	($\frac{1}{5}$ =) 0.2	($\frac{1}{2}$ =) 0.5	($\frac{1}{1}$ =) 1

Since **d∝$\frac{1}{V}$** or **d = $\frac{k}{V}$** we've added one row, $\frac{1}{V}$, to our table (shown in red). A graph of **d** against $\frac{1}{V}$ will give us a straight line.

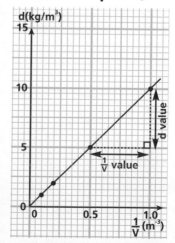

gradient = k

$= \frac{d \text{ value}}{\frac{1}{V} \text{ value}}$

$= \frac{(10 - 5)}{(1 - 0.5)}$

$= \frac{5}{0.5}$

$= 10$

Relationship is:

d = $\frac{10}{V}$

Angles 1

All angles are measured in DEGREES (°). A protractor can be used to measure the size of an angle.

Acute, Right, Obtuse and Reflex Angles

An angle LESS THAN 90° is called an **ACUTE ANGLE**

An angle EQUAL TO 90° is called a **RIGHT ANGLE**

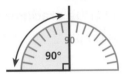

An angle GREATER THAN 90° but LESS THAN 180° is called an **OBTUSE ANGLE**

An angle GREATER THAN 180° is called a **REFLEX ANGLE**

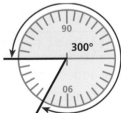

Angles on a Straight Line, at a Point and Vertically Opposite

Angles on a straight line add up to 180° (also known as Adjacent Angles)

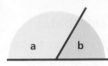

a = 120°
b = 60°   a + b = 180°

p = 45°
q = 90°   p + q + r = 180°
r = 45°

Angles at a point add up to 360°

a = 240°
b = 120°   a + b = 360°

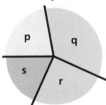

p = 80°
q = 125°
r = 90°   p + q + r + s = 360°
s = 65°

Vertically opposite angles are equal

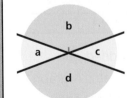

a = 40°
c = 40°   } Vertically opposite
b = 140°
d = 140°   } Vertically opposite

Examples...

1

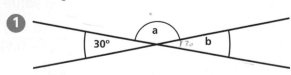

a = 180° − 30° = 150°
(Angles on a straight line add up to 180°)
b = 30° (Vertically opposite angles are equal)

2 c = 360° − (120° + 160°)
c = 360° − 280° = 80°
(Angles at a point add up to 360°)

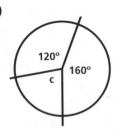

Angles 2

Parallel Lines

Parallel lines run in exactly the same direction and never meet. Parallel lines are shown by arrows. There is no limit to the number of lines which may run parallel to each other. When a straight line crosses two or more parallel lines, corresponding, alternate and allied angles are formed.

Alternate Angles

- Alternate angles are formed on opposite (alternate) sides of a line which crosses two or more parallel lines.
- Alternate angles are always equal in size.
- Alternate angles can be easily spotted because they form a letter **Z** (although sometimes it may be reversed, **Ƨ**).

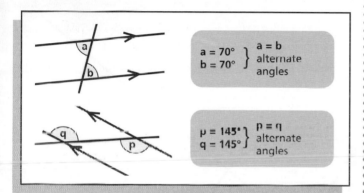

$a = 70°$
$b = 70°$ } $a = b$ alternate angles

$p = 145°$
$q = 145°$ } $p = q$ alternate angles

Corresponding Angles

- Corresponding angles are formed on the same side of a line, which crosses two or more parallel lines. They all appear in matching (corresponding) positions above or below the parallel lines.
- Corresponding angles are always equal in size.
- Corresponding angles can be easily spotted because they form a letter **F** (although sometimes it may be reversed, **ꟻ**, or upside down **Ⅎ,ꟻ**).

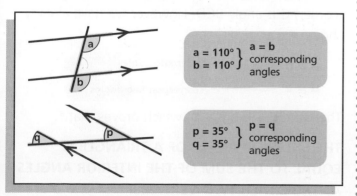

$a = 110°$
$b = 110°$ } $a = b$ corresponding angles

$p = 35°$
$q = 35°$ } $p = q$ corresponding angles

Allied (Co-Interior) Angles

- Allied angles are formed on the same side of a line, which crosses two or more parallel lines. They appear inside two parallel lines, facing each other.
- Allied angles always add up to 180°.
- Allied angles can be easily spotted because they form a letter **⊏** or **⊔** (although sometimes it may be reversed, **⊐** or **⊓**).

* Allied angles do not appear on the course specification. They are included here because they will help you to understand the relationships between the different angles formed when a straight line crosses two or more parallel lines. You will find them useful when calculating angles too.

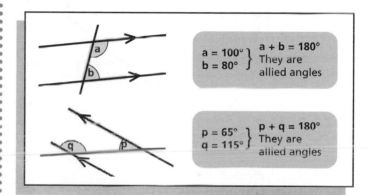

$a = 100°$
$b = 80°$ } $a + b = 180°$ They are allied angles

$p = 65°$
$q = 115°$ } $p + q = 180°$ They are allied angles

Example

Calculate the angles **a**, **b**, and **c** in relation to **x** in the following parallelogram.

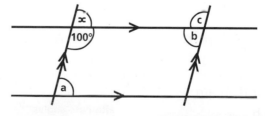

$x + 100° = 180°$ ← angles on a straight line
$\therefore x = 80°$

$a = x$ ← corresponding angles
$\therefore a = 80°$

$b = x$ ← alternate angles
$\therefore b = 80°$

$c + x = 180°$ ← allied angles
$\therefore c = 100°$

Triangles

A triangle is a 3-sided two-dimensional shape. The interior angles of the triangle below are **a**, **b** and **c**.

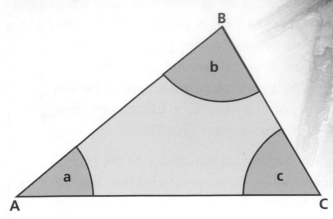

If we extend the side **AC** to point **D**, and add a line from **C** to **E** which runs parallel to **AB** then we get the following diagram.

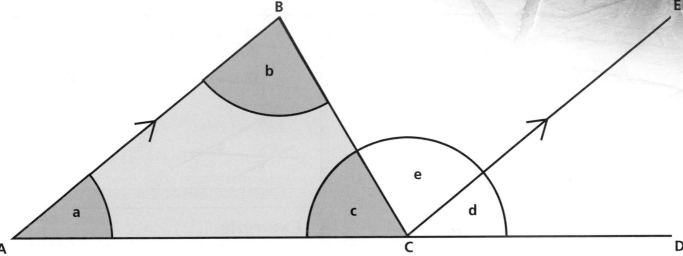

We can now say the following things about these particular angles:

❶ The Interior Angles of a Triangle add up to 180°

From our diagram...

e = b ← alternate angles

d = a ← corresponding angles

However...

c + e + d = 180° ← angles on a straight line add up to 180°

Therefore...

c + b + a = 180°, which proves that...

THE INTERIOR ANGLES OF A TRIANGLE ADD UP TO 180°

❷ The Exterior Angle of a Triangle is equal to the sum of the Interior Angles at the other two vertices

The exterior angle of this triangle at the vertex (corner) C is angle BCD. However we have already shown that:

e = b ← alternate angles

d = a ← corresponding angles

Therefore **d + e = a + b**, which proves that...

THE EXTERIOR ANGLE OF A TRIANGLE IS EQUAL TO THE SUM OF THE INTERIOR ANGLES AT THE OTHER TWO VERTICES.

Types of Triangle

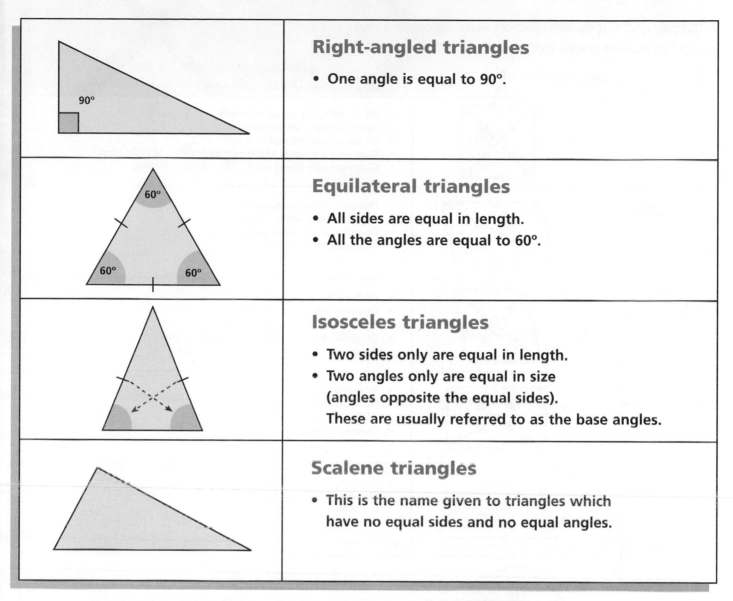

Right-angled triangles

- One angle is equal to 90°.

Equilateral triangles

- All sides are equal in length.
- All the angles are equal to 60°.

Isosceles triangles

- Two sides only are equal in length.
- Two angles only are equal in size
  (angles opposite the equal sides).
  These are usually referred to as the base angles.

Scalene triangles

- This is the name given to triangles which
  have no equal sides and no equal angles.

Example...

The following diagram shows an isosceles and a
right-angled triangle. Calculate the angles **a** to **f**
explaining your reasoning.

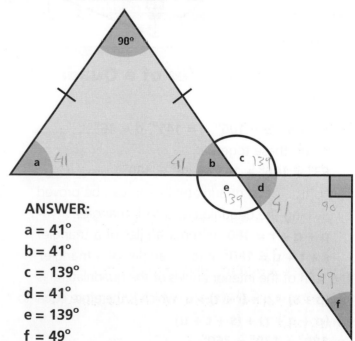

a = b (Base angles of an isosceles triangle)

But, **a + b + 98° = 180°** (Interior angles of a triangle)

∴ **a + b = 82°** so each one is equal to **41°**.

b + c = 180° (Angles on a straight line)

c = 180° − 41° = 139°

b = d (Vertically opposite angles) ∴ **d = 41°**

c = e (Vertically opposite angles) ∴ **e = 139°**

d + f + right-angle = 180°

(Interior angles of a triangle)

∴ **f = 180° − 90° − 41° = 49°**

ANSWER:

a = 41°

b = 41°

c = 139°

d = 41°

e = 139°

f = 49°

A quadrilateral is a 4-sided, two-dimensional shape which has interior angles that add up to 360°.

Types of Quadrilateral

Square		• All the sides are equal in length • Opposite sides are parallel • All the angles are equal to 90° • Diagonals are equal and bisect each other at right-angles. • Diagonals also bisect each of the interior angles.
Parallelogram		• Opposite sides are equal in length • Opposite sides are parallel • Opposite angles are equal in size • Diagonals bisect each other.
Rhombus		• All the sides are equal in length • Opposite sides are parallel • Opposite angles are equal in size • Diagonals bisect each other at right-angles • Diagonals also bisect the interior angles.
Rectangle		• Opposite sides are equal in length • Opposite sides are parallel • All the angles are equal to 90° • Diagonals are equal and bisect each other.
Trapezium		• No sides equal in length • One pair of sides parallel • No angles equal in size
Kite		• 2 pairs of equal adjacent sides • 1 pair of opposite equal angles • Diagonals cross at right-angles and one bisects the other.

The Interior Angles of a Quadrilateral

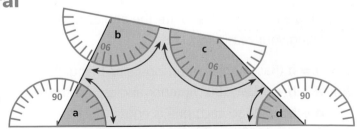

In this quadrilateral…

$a = 63°$, $b = 106°$, $c = 145°$, $d = 46°$

If we add these together…

$63° + 106° + 145° + 46° = 360°$

This is true of all quadrilaterals and can be proved by dividing the quadrilateral into 2 triangles:

$p + q + r = 180°$ (Interior angles of a triangle)

$s + t + u = 180°$ (Interior angles of a triangle)

The sum of the interior angles of the quadrilateral is

$(p + s) + q + (r + t) + u$ which is therefore equal to

$(p + q + r) + (s + t + u)$

i.e. $180° + 180° = 360°$

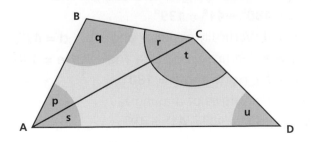

Irregular Polygons

Polygons

A polygon is a two-dimensional shape with 3 or more sides. We have already seen the 3-sided polygon (triangle) and the 4-sided polygon (quadrilateral). A polygon is said to be REGULAR if all its sides and all its angles are equal. Otherwise it is known as an IRREGULAR polygon. Take a look at the two irregular polygons below.

Interior and Exterior Angles of a Polygon

The angles inside a polygon are called the INTERIOR ANGLES and those outside are called the EXTERIOR ANGLES. As for triangles and quadrilaterals the size of each of these angles can be measured using a protractor.

Interior angles	Exterior angles	Interior + Exterior angles
a = 100°	p = 80°	a + p = 180°
b = 100°	q = 80°	b + q = 180°
c = 140°	r = 40°	c + r = 180°
d = 70°	s = 110°	d + s = 180°
e = 130°	t = 50°	e + t = 180°
a+b+c+d+e = 540°	p+q+r+s+t = 360°	

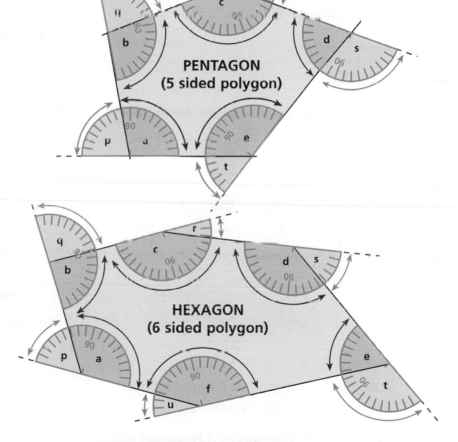

PENTAGON (5 sided polygon)

Interior angles	Exterior angles	Interior + Exterior angles
a = 121°	p = 59°	a + p = 180°
b = 89°	q = 91°	b + q = 180°
c = 158°	r = 22°	c + r = 180°
d = 136°	s = 44°	d + s = 180°
e = 63°	t = 117°	e + t = 180°
f = 153°	u = 27°	f + u = 180°
a+b+c+d+e+f = 720°	p+q+r+s+t+u = 360°	

HEXAGON (6 sided polygon)

We can see from above that…

1. **The Exterior Angles of a polygon always add up to 360°**
2. **The Interior Angle + the Exterior Angle always add up to 180°**

We can also see that the interior angles of different polygons do not add up to the same number of degrees. A triangle is 180°, a quadrilateral is 360°, a pentagon is 540°, a hexagon is 720°. The more sides the polygon has the greater the sum of the interior angles.

The sum of the interior angles
= (n – 2) x 180°
(where n = number of sides)

Name of Polygon	Number of sides (n)	Sum of Interior Angles, (n – 2) x 180°	Sum of Exterior Angles
Triangle	3	180°	360°
Quadrilateral	4	360°	360°
Pentagon	5	540°	360°
Hexagon	6	720°	360°

Regular Polygons

Regular Polygons

These are polygons that have…

- **Sides of the same length**
- **Interior angles of the same size**
- **Exterior angles of the same size**

Here are four examples of regular polygons:

Knowing that the exterior angles of a polygon add up to 360°, that the interior angles are equal in a regular polygon and that the exterior and interior angles add up to 180° enables various calculations to be performed:

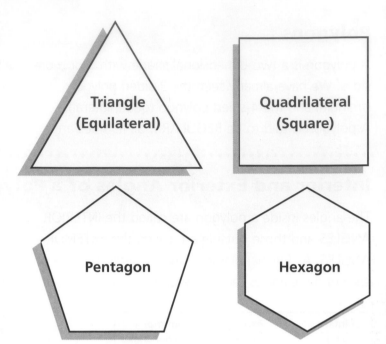

Triangle (Equilateral)

Quadrilateral (Square)

Pentagon

Hexagon

Examples...

1 Calculate the size of **a)** each exterior, and **b)** each interior angle for a regular hexagon (6 sides).

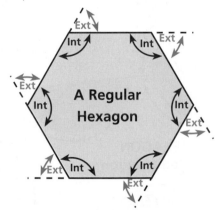

A Regular Hexagon

A Regular Hexagon has 6 equal Interior angles and 6 equal Exterior angles.

a) The Exterior angles of a Hexagon add up to 360° (see previous page)

Each Exterior Angle = $\frac{360°}{6}$ = 60°

b) The Interior angle + the Exterior angle add up to 180° (see previous page)

Each Interior Angle
= 180° – Exterior Angle
= 180° – 60° = 120°

2 A regular polygon has each interior angle = 108°. Calculate **a)** the size of each exterior angle, and **b)** the number of sides the polygon has.

a) The Interior angle + the Exterior angle add up to 180°.

Each Exterior Angle
= 180° – Interior Angle
= 180° – 108° = 72°

b) The Exterior angles of a Polygon add up to 360° (see previous page)

Number of Exterior Angles = $\frac{360°}{72°}$ = 5

Number of sides = 5
(i.e. it is a regular PENTAGON)

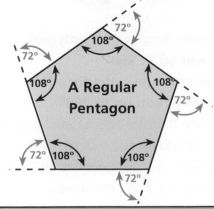

A Regular Pentagon

Congruence and Tessellation

Congruent Shapes

These boys are IDENTICAL in their SIZE and SHAPE although their position relative to each other may be different. These four boys are congruent:

These four shapes are congruent:

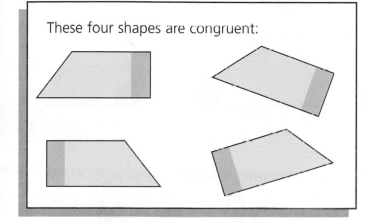

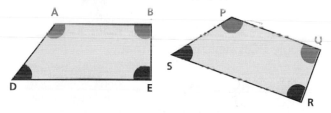

Shape ABCD and shape PQRS are CONGRUENT because $\hat{A} = \hat{P}$, $\hat{B} = \hat{Q}$, $\hat{C} = \hat{R}$ and $\hat{D} = \hat{S}$

TWO CONGRUENT SHAPES HAVE ANGLES THE SAME SIZE and also AB = PQ, BC = QR, CD = RS and DA = SP

TWO CONGRUENT SHAPES HAVE SIDES OF THE SAME LENGTH

Congruent Triangles

Two triangles are congruent if they both satisfy the following conditions:

1 Three Sides, **SSS**.

AB = PQ
BC = QR
AC = PR
∴ **ABC** and **PQR** are congruent triangles.

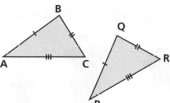

2 Two Sides and Included Angle, **SAS**.

AB = PQ
AC = PR
$\hat{A} = \hat{P}$
∴ **ABC** and **PQR** are congruent triangles.

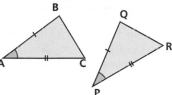

3 Two Angles and One Corresponding Side, **ASA**.

AB = PQ
$\hat{A} = \hat{P}$
$\hat{B} = \hat{Q}$
∴ **ABC** and **PQR** are congruent triangles.

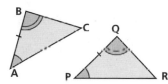

Note: In our diagram we've shown AB = PQ. However we could have shown BC = QR or AC = PR. The two triangles are still congruent.

4 Right Angle, Hypotenuse and One Other Side, **RHS**.

AB = PQ
BC = QR
$\hat{C} = \hat{R} = 90°$
∴ **ABC** and **PQR** are congruent triangles.

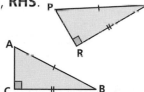

Note: In our diagram we've shown BC = QR. However we could have shown BC = PR, AC = QR or AC = PR. The two triangles would still be congruent.

Tessellations

A TESSELLATION is a pattern of CONGRUENT SHAPES that fit together with NO GAPS IN BETWEEN to cover a flat surface. Not all congruent shapes form a tessellation. A tessellation can also be made using two or more congruent shapes.

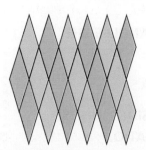

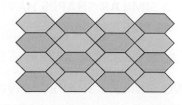

Similarity

Similar Shapes

These boys are IDENTICAL in their SHAPE but they are NOT IDENTICAL in SIZE (they can be bigger or smaller). Yet again their position relative to each other may be different. These four boys are SIMILAR:

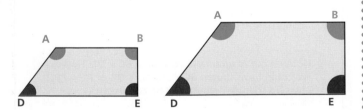

These four shapes are similar.

This is our original shape

This is our original shape, but bigger in size.

This is our original shape, but smaller in size.

This is our original shape, smaller in size, upside down and turned slightly clockwise.

Shape ABCD and shape PQRS are SIMILAR because $\hat{A} = \hat{P}$, $\hat{B} = \hat{Q}$, $\hat{C} = \hat{R}$ and $\hat{D} = \hat{S}$

TWO SIMILAR SHAPES HAVE ANGLES THE SAME SIZE and also $\frac{AB}{PQ} = \frac{BC}{QR} = \frac{CD}{RS} = \frac{DA}{SP}$

or AB:PQ = BC:QR = CD:RS = DA:SP

TWO SIMILAR SHAPES HAVE SIDES WHOSE LENGTHS ARE IN THE SAME RATIO.

Example

The two triangles below are similar.
Calculate the length of…
a) QR
b) AC

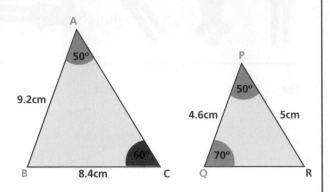

In triangle **ABC**, $\hat{B} = 70°$ $(180° - (60° + 50°))$ and in triangle **PQR**, $\hat{R} = 60°$ $(180° - (70° + 50°))$ And so $\hat{A} = \hat{P}$, $\hat{B} = \hat{Q}$, $\hat{C} = \hat{R}$

This means that triangle **ABC** is similar to triangle **PQR** which means that…

$$\frac{AB}{PQ} = \frac{BC}{QR} = \frac{AC}{PR}$$

> Don't assume that the shapes (in this case triangles) are always lettered in alphabetical order

$$\frac{9.2cm}{4.6cm} = \frac{8.4cm}{QR} = \frac{AC}{5cm}$$

We can now calculate the unknown lengths…

a) $\frac{9.2cm}{4.6cm} = \frac{8.4cm}{QR}$

$QR = \frac{8.4cm \times 4.6cm}{9.2cm}$

> Rearranged to get QR on its own

$QR = 4.2cm$

b) $\frac{9.2cm}{4.6cm} = \frac{AC}{5cm}$

$AC = \frac{9.2cm \times 5cm}{4.6cm}$

> Rearranged to get AC on its own

$AC = 10cm$

Pythagoras' Theorem 1

Pythagoras

Pythagoras was a Greek philosopher and mathematician who lived over 2000 years ago. His theorem is used to calculate the length of an unknown side in a right-angled triangle when the lengths of the other two sides are known. The theorem states… **'the square on the Hypotenuse of a right-angled triangle is equal to the sum of the squares on the other two sides'.**

This is shown in our diagram. The square on the hypotenuse (the longest side, always found opposite the right-angle) is 25, which is the sum (16 + 9) of the squares on the other two sides. This can be summarised using the formula $c^2 = a^2 + b^2$. Often, you are asked to work out the length of the hypotenuse (c). In our diagram, we know that $c^2 = 25$ **squares**. So, if you remember powers and roots (pages 13 - 15)… $c^2 = 25$ and so $c = \sqrt{25} = 5$ **units**

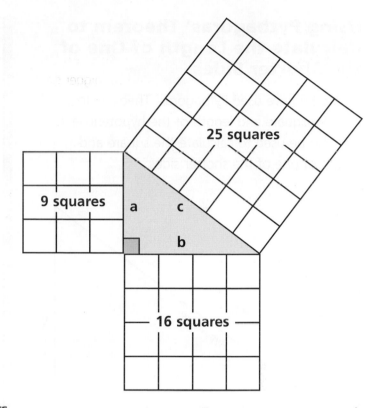

Examples…

1 Calculate the length of **c** in the following right-angled triangle.

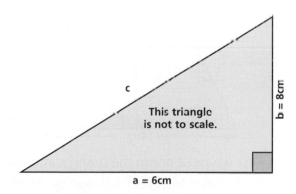

a = 6cm

Using Pythagoras' Theorem

$c^2 = a^2 + b^2$

$c^2 = 6^2 + 8^2$

$c^2 = 36 + 64$

$c^2 = 100$

To get **c** we need to take the square root.

$c = \sqrt{100}$

$c = 10$cm (remember the units)

2 Calculate the length of **c** in the following right-angled triangle to 1 decimal place.

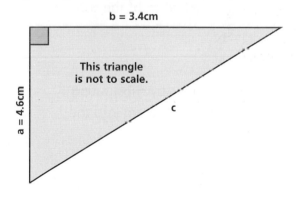

b = 3.4cm

Using Pythagoras' Theorem

$c^2 = a^2 + b^2$

$c^2 = 4.6^2 + 3.4^2$

$c^2 = 21.16 + 11.56$

$c^2 = 32.72$

To get **c** we need to take the square root.

$c = \sqrt{32.72}$

$c = 5.7$cm (remember the units)

Pythagoras' Theorem 2

Using Pythagoras' Theorem to Calculate the Length of One of the Shorter Sides

So far we have used Pythagoras' Theorem to find the square and length of the hypotenuse. It can also be used to calculate the square and length of one of the shorter sides.

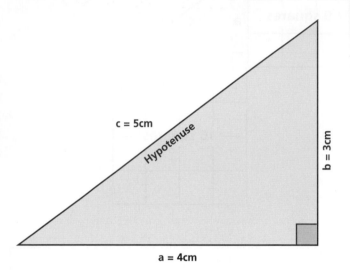

If we look at our formula $c^2 = a^2 + b^2$, it can be rearranged to make a^2 or b^2 the subject:

$$a^2 = c^2 - b^2$$
$$b^2 = c^2 - a^2$$

We can check this works by substituting the side lengths on our diagram into the rearranged formula.

$$a^2 = c^2 - b^2$$
$$4^2 = 5^2 - 3^2$$
$$16 = 25 - 9 \checkmark$$

$$b^2 = c^2 - a^2$$
$$3^2 = 5^2 - 4^2$$
$$9 = 25 - 16 \checkmark$$

The square on one of the shorter sides	=	The square on the Hypotenuse	−	The square on the other short side

Examples...

1 Calculate the length of **b** in the following right-angled triangle.

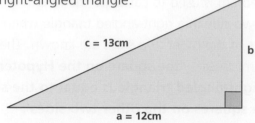

Using Pythagoras' Theorem rearranged.

$$b^2 = c^2 - a^2$$
$$b^2 = 13^2 - 12^2$$
$$b^2 = 169 - 144$$
$$b^2 = 25$$

To get **b** we need to take the square root

$$b = \sqrt{25} = 5cm$$

2 Calculate the height of the isosceles triangle shown below, to 3 significant figures.

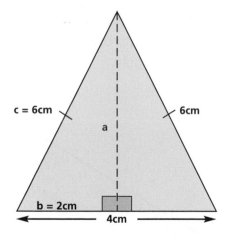

First we must divide the triangle into two right-angled triangles (as shown by the dotted line) in order to use Pythagoras' theorem and label the one we are going to work with (in red).

$$a^2 = c^2 - b^2$$ ← This would be true for either right-angled triangle
$$a^2 = 6^2 - 2^2$$
$$a^2 = 36 - 4$$ ← half the base of the isosceles triangle
$$a^2 = 32$$

To get **a** we need to take the square root

$$a = \sqrt{32} = 5.66cm$$

Trigonometry

The three trigonometrical ratios **sin**, **cos** and **tan** (short for **sine**, **cosine** and **tangent**) can only be used with right-angled triangles (just like Pythagoras' Theorem). Each trigonometrical ratio involves ONE ANGLE and TWO SIDES and can be used to calculate the length of an unknown side or the value of an unknown angle. To use any of the trig. ratios you must first identify and label the different parts of your triangle.

The three sides are…

- **HYPOTENUSE**… this is the longest side, **AB**, opposite the right angle.
- **OPPOSITE**… this is the side, **BC**, opposite the angle θ.
- **ADJACENT**… this is the side, **AC**, next to (adjacent to) the angle θ, although if you have identified the opposite and the hypotenuse this is obviously the remaining side.

θ (theta) is a Greek letter commonly used for unknown angles. We could use x instead.

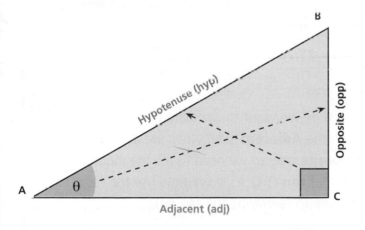

The three trigonometrical ratios are…

$$\sin\theta = \frac{opp}{hyp} \qquad \cos\theta = \frac{adj}{hyp} \qquad \tan\theta = \frac{opp}{adj}$$

This very conveniently forms the name of a Native American Indian chief **'SOH CAH TOA'**. Remember this and you'll never confuse these ratios.

Using a Formula Triangle

Each of the trigonometrical ratios can be put into a formula triangle. This will make things easier when you have to rearrange a ratio. The three formula triangles are…

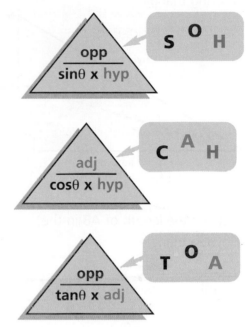

Let's take the sin ratio, SOH…
To rearrange the formula to find the **opposite**, cover the **opp** in the triangle with your finger

… to give
opp = sinθ x hyp

To rearrange the formula to find the **hypotenuse**, cover the **hyp** in the triangle with your finger

… to give **hyp =** $\dfrac{opp}{sin\theta}$

Remember, you can do the same with the other two ratios. Make sure that you practise rearranging them.

Trigonometry 2

Using Trigonometry to Calculate the Length of an Unknown Side

Examples...

❶ Calculate the length of BC in the following triangle.

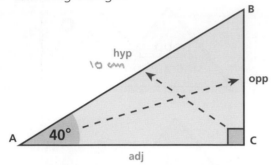

We want to calculate the length of the **Opposite** side and we are given the **Hypotenuse**, so we need to use the trigonometrical ratio for **sin** (S O H). If we now use the formula triangle to rearrange the sin ratio:

opp = sinθ x hyp

  BC = sin40° x 10cm

  BC = 0.643 x 10cm

  BC = 6.4cm (to 1 d.p.)

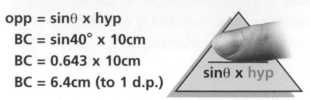

❷ Calculate the length of AB in the following triangle.

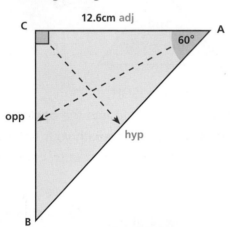

This time we want to calculate the length of the **Hypotenuse** and we are given the **Adjacent**, so we need to use the trigonometrical ratio for **cos** (C A H). If we now use the formula triangle to rearrange the cos ratio:

$$hyp = \frac{adj}{cos\theta}$$

$$AB = \frac{12.6cm}{cos\ 60°}$$

$$AB = \frac{12.6}{0.5}$$

AB = 25.2cm

❸ A ladder leans against a vertical wall as shown below. The foot of the ladder is 1.6m from the wall and the ladder makes an angle of 30° with the wall. Calculate the height of the wall.

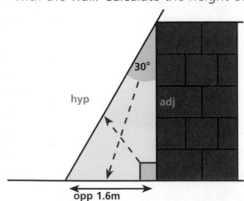

This time we want to calculate the height of the wall (the **Adjacent** side) when we are given the **Opposite** side, so we need to use the trigonometrical ratio for **tan** (T O A). If we now use the formula triangle to rearrange the tan ratio:

$$adj = \frac{opp}{tan\theta}$$

$$adj = \frac{1.6m}{tan\ 30°}$$

$$adj = \frac{1.6}{0.577}$$

height of wall = 2.77m (to 2 d.p.)

Using Trigonometry to Calculate the Size of an Unknown Angle

Example

Calculate the size of angle BAC in the following triangle.

Since we are given the **Adjacent** and **Hypotenuse**, we need to use the trig. ratio for **cos** (C A H).

$$\cos \theta = \frac{Adj}{Hyp}$$

> There is no need to use a formula triangle to calculate the size of an unknown angle. However you still need to remember 'SOH CAH TOA'

$$\cos B\hat{A}C = \frac{6cm}{12cm}$$

$$\cos B\hat{A}C = 0.5$$

$$B\hat{A}C = \cos^{-1}0.5$$

$$B\hat{A}C = 60°$$

> To calculate BÂC we now have to use the INVERSE cos button on our calculator, i.e. the cos⁻¹ button to get the answer

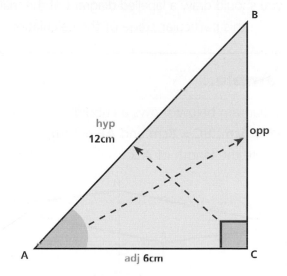

hyp 12cm

opp

adj 6cm

A

B

C

Angles of Elevation and Depression

Both of these are angles measured from the horizontal direction either upwards or downwards.

A child looking up at an adult...

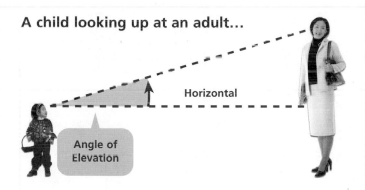

Horizontal

Angle of Elevation

An adult looking down at a child...

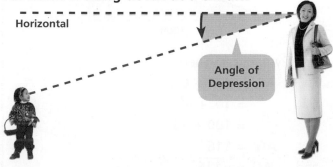

Horizontal

Angle of Depression

These two angles will always be equal.

Example

The tree in the diagram below is 9m high. Calculate the angle of elevation to the top of the tree from a point on the ground 20m from the foot of the tree.

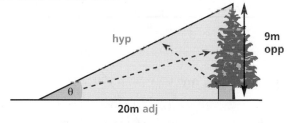

hyp

9m opp

θ

20m adj

We want to calculate the angle of elevation (θ) and we are given the **Opposite** and **Adjacent**, so we need to use the trig. ratio for **tan** (T O A).

$$\tan \theta = \frac{Opposite}{Adjacent}$$

$$\tan \theta = \frac{9m}{20m}$$

$$\tan \theta = 0.45$$

$$\theta = \tan^{-1} 0.45$$

> To calculate the angle of elevation θ, we now have to use the INVERSE tan button on our calculator i.e. the tan⁻¹ button to get the answer

Angle of Elevation = 24.2° (to 1 d.p.)

Problems Involving 3-D Figures 1

Problems involving 3-D figures very often have to be broken down into two (or more) stages. At each stage you should draw a labelled diagram of the triangle involved in this particular stage of the calculation.

Example...

The diagram below shows a cuboid.

If **AB = 6cm**, **BC = 8cm** and **AE = 4cm**, calculate the length of **AG**.

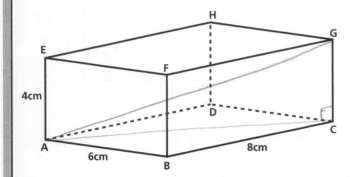

Since our figure is a cuboid then...

AB = DC = EF = HG = 6cm

BC = AD = FG = EH = 8cm

AE = BF = CG = DH = 4cm

To calculate the length of **AG** we need to do it in two stages. **AG** is one side of triangle **ACG**, with $\hat{C} = 90°$. Side **AC** is unknown, however **AC** is also one side of triangle **ABC**, with $\hat{B} = 90°$, and both **AB** and **BC** are known.

STAGE 1

Calculation of **AC**.

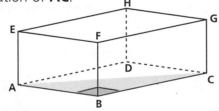

If we take triangle **ABC** with $\hat{B} = 90°$:

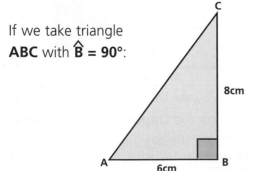

Using Pythagoras' Theorem...

$$AC^2 = AB^2 + BC^2$$
$$= 6^2 + 8^2$$
$$= 36 + 64$$
$$AC^2 = 100$$

and so **AC** = $\sqrt{100}$

∴ **AC = 10cm**

NOTE:
We could also have used triangle **ADC** to find **AC**.

STAGE 2

Calculation of **AG**.

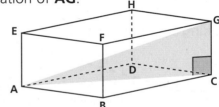

If we take triangle **ACG** with $\hat{C} = 90°$:

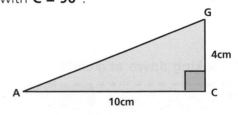

Using Pythagoras' Theorem...

$$AG^2 = AC^2 + CG^2$$
$$= 10^2 + 4^2$$
$$= 100 + 16$$
$$AG^2 = 116$$

and so **AG** = $\sqrt{116}$

∴ **AG = 10.77cm (2 d.p.)**

Problems Involving 3-D Figures 2

Calculation of the Angle between a Line and a Plane

ABCD is a horizontal plane and **PQ** is a line where **Q** is directly above **R**, a point on **DC**.

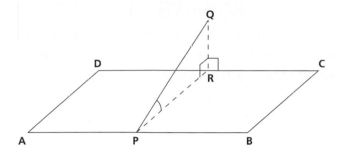

To determine the angle between the line **PQ** and the plane **ABCD**:

1 One side of the triangle that includes the angle must lie on the plane itself. **PR** lies on the plane **ABCD**.

2 The triangle which includes the angle must also be perpendicular (i.e. at right-angles) to the plane. Triangle **PRQ** is perpendicular to the plane **ABCD**.

The angle between the line **PQ** and the plane **ABCD** is **QP̂R**, which can now be calculated.

Example...

The diagram below shows a square-based pyramid where **AB = 7cm** and **AE = BE = CE = DE = 12cm**. **E** lies directly above **X**. Calculate the angle between the line **AE** and the base **ABCD**.

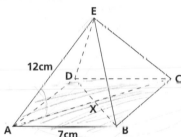

The angle between **AE** and the base **ABCD** (a horizontal plane) is **EÂX**, since side **AX** lies on the base and triangle **EAX** is perpendicular to the base. To calculate **EÂX** we need to do it in two stages. **EÂX** is one angle in triangle **EAX**, with **X̂ = 90°**. Both **EX** and **AX** are unknown, however **AX** is half of **AC** and **AC** is one side of triangle **ABC**, with **B̂ = 90°** and **AB** and **BC** are both known.

STAGE 1

Calculation of **AC**.

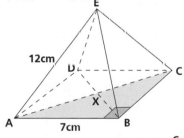

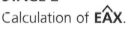

Using Pythagoras' Theorem...

$$AC^2 = AB^2 + BC^2$$
$$= 7^2 + 7^2$$
$$= 49 + 49$$
$$AC^2 = 98$$

and so $AC = \sqrt{98}$

$$\therefore AC = 9.90 \text{ (2 d.p.)}$$

And consequently...

$$AX = \tfrac{1}{2} AC = \tfrac{1}{2} \times 9.90\text{cm} = 4.95\text{cm}$$

STAGE 2

Calculation of **EÂX**.

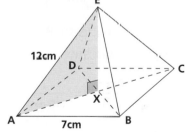

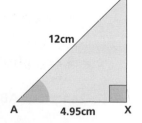

Using the trigonometrical ratio for **cos** (CAH)...

$$\cos E\hat{A}X = \frac{Adj}{Hyp}$$
$$= \frac{AX}{AE}$$
$$= \frac{4.95}{12}$$

$$\cos E\hat{A}X = 0.4125$$
$$\therefore E\hat{A}X = \cos^{-1}(0.4125)$$
$$= 65.6° \text{ (to 1 d.p.)}$$

The Sine Rule

The sine rule can be used in **any** triangle to calculate the length of an unknown side or the size of an unknown angle. In any triangle…

$$\frac{a}{\sin A} = \frac{b}{\sin B} = \frac{c}{\sin C}$$

where **a**, **b** and **c** are the sides opposite angles **A**, **B** and **C** respectively.

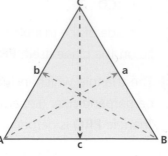

You can only use two parts of the sine rule at any particular time:

$$\frac{a}{\sin A} = \frac{b}{\sin B} \quad \text{or} \quad \frac{a}{\sin A} = \frac{c}{\sin C} \quad \text{or} \quad \frac{b}{\sin B} = \frac{c}{\sin C}$$

The sine rule can be used to calculate…
* The length of an unknown side if we are given the length of one other side and the size of two angles.
* The size of an unknown angle if we are given the length of any two sides and the size of an angle opposite one of these sides.

Examples…

1 Calculate the length of **AB** in the following triangle:

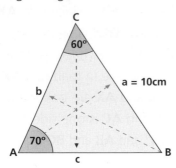

Using the sine rule…

$$\frac{a}{\sin A} = \left[\frac{b}{\sin B}\right] = \frac{c}{\sin C}$$

Since **b** and **sin B** play no part in the question - ignore them.

$$\therefore \frac{10\text{cm}}{\sin 70°} = \frac{c}{\sin 60°}$$

Rearranging gives us…

$$c = \frac{10 \times \sin 60°}{\sin 70°}$$

$$c = 9.22\text{cm (to 2 d.p)}$$

$\therefore$ Length of **AB = 9.22cm**

2 Calculate the size of **AB̂C** in the following triangle:

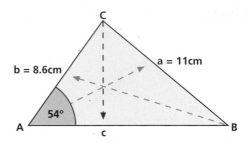

Using the sine rule…

$$\frac{a}{\sin A} = \frac{b}{\sin B} = \left[\frac{c}{\sin C}\right]$$

Since **c** and **sin C** play no part in the question - ignore them .

$$\therefore \frac{11\text{cm}}{\sin 54°} = \frac{8.6\text{cm}}{\sin B}$$

Rearranging gives us…

$$\sin B = \frac{8.6 \times \sin 54°}{11} = 0.6325$$

$$\hat{B} = \sin^{-1}(0.6325) = 39.23° \text{ (to 2 d.p)}$$

Size of **AB̂C = 39.23°**

Note: The sine rule can be used to solve problems in 2-D or 3-D. Also it can be used with right-angled triangles **but** you should always use the three trig. ratios **sin, cos** and **tan** as they are a lot easier.

The cosine rule, like the sine rule, can be used in **any** triangle to calculate the length of an unknown side or the size of an unknown angle. In any triangle…

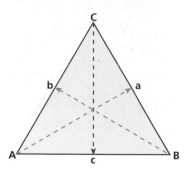

$$a^2 = b^2 + c^2 - 2bc \cos A$$

… to find the size of an unknown side.

And if we rearrange:

$$\cos A = \frac{b^2 + c^2 - a^2}{2bc}$$

… to find the size of an unknown angle.
If we wanted to use the cosine rule to find the length of **b** or **c** then it becomes:

$$b^2 = a^2 + c^2 - 2ac \cos B$$
$$c^2 = a^2 + b^2 - 2ab \cos C$$

Also for angles **B** and **C** it becomes:

$$\cos B = \frac{a^2 + c^2 - b^2}{2ac}$$

$$\cos C = \frac{a^2 + b^2 - c^2}{2ab}$$

The cosine rule (compare to the sine rule) can be used to calculate…

- The length of an unknown side if we are given the length of two sides and the size of their included angle.
- The size of an unknown angle if we are given the length of all three sides.

Note: Like the sine rule, the cosine rule can be used to solve problems in 2-D or 3-D. Also it can be used with right-angled triangles **but** you should always use the three trig. ratios sin, cos and tan as they are a lot easier.

Examples…

① Calculate the length of **BC** in the following triangle:

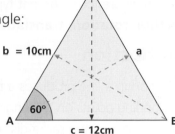

Using the cosine rule:

$$a^2 = b^2 + c^2 - 2bc \cos A$$
$$= 10^2 + 12^2 - (2 \times 10 \times 12 \times \cos 60°)$$
$$= 100 + 144 - (2 \times 10 \times 12 \times 0.5)$$
$$= 100 + 144 - 120$$
$$= 124$$
$$a = \sqrt{124} = 11.14\text{cm (to 2 d.p.)}$$

∴ Length of **BC** = 11.14cm

It would be impossible for you to calculate the length of **BC** using the sine rule - try it!

② Calculate the size of **BÂC** in the following triangle:

Using the cosine rule (rearranged):

$$\cos A = \frac{b^2 + c^2 - a^2}{2bc}$$

$$= \frac{9.1^2 + 7.6^2 - 8.4^2}{2 \times 9.1 \times 7.6}$$

$$= \frac{82.81 + 57.76 - 70.56}{138.32}$$

$$\cos A = 0.5061$$
$$\hat{A} = \cos^{-1}(0.5061) = 59.60° \text{ (to 2 d.p.)}$$

∴ size of **BÂC** = 59.60°

Again it would be impossible for you to calculate the size of **BÂC** using the sine rule - try it!

Transformations 1

Types of Transformation

A transformation is a process which changes the position (and possibly the size and orientation) of a shape. There are four different types of transformation: **reflection**, **rotation**, **translation**, **enlargement**.

① Reflection

A reflection in a line produces a mirror image in which corresponding points on the original shape and the mirror image are always the same distance from the mirror line. A line joining corresponding points always crosses the mirror line at 90°. To describe a reflection you must specify the MIRROR LINE by giving the equation of the line of reflection (in the examples on the right these are $x = 6$ and $y = 1$). As you can see, in a reflection, the ORIENTATION and POSITION changes but everything else stays the same, i.e. length and angle. The object and image are congruent shapes (see page 79). Difficult reflections can be completed more easily if you use tracing paper. The example below is a reflection in the line $y = x$.

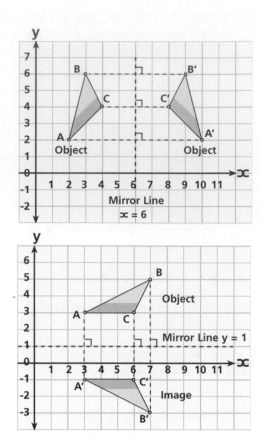

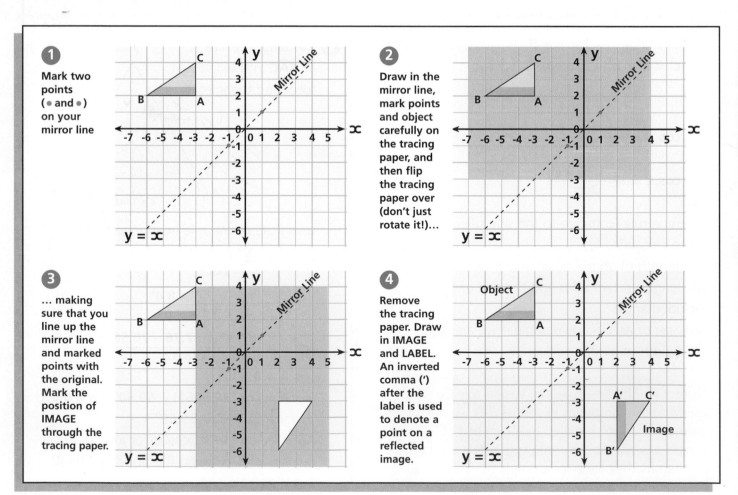

① Mark two points (● and ●) on your mirror line

② Draw in the mirror line, mark points and object carefully on the tracing paper, and then flip the tracing paper over (don't just rotate it!)...

③ ... making sure that you line up the mirror line and marked points with the original. Mark the position of IMAGE through the tracing paper.

④ Remove the tracing paper. Draw in IMAGE and LABEL. An inverted comma (') after the label is used to denote a point on a reflected image.

Transformations 2

❷ Rotation

A rotation turns a shape through a clockwise or anti-clockwise angle about a fixed point known as the Centre of Rotation. All lines in the shape rotate through the same angle. Rotation, (just like reflection) changes the ORIENTATION and POSITION of the shape, but everything else stays the same, i.e. length

and angle. The object and image are congruent shapes (see page 79). To describe a rotation, you must specify the following three things:
- The DIRECTION OF TURN (clockwise/anti-clockwise)
- The CENTRE OF ROTATION
- The ANGLE TURNED.

Examples...

Rotation of 90° clockwise about the origin (0,0)

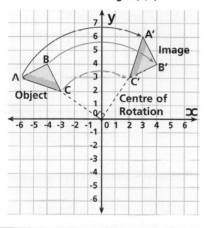

Rotation of 180° clockwise about the origin (0,0)

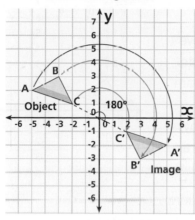

Rotation of 270° clockwise about the origin (0,0) (is the same as a rotation of 90° anti-clockwise)

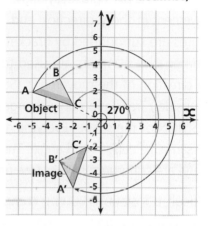

Difficult rotations can be completed more easily if you use tracing paper. If we wanted to rotate triangle ABC 90° clockwise about the centre (-1,1)...

❶ Draw in axes and object carefully on the tracing paper

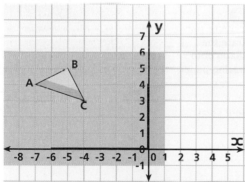

❷ Place point of compass on centre of rotation (-1,1) and rotate paper clockwise

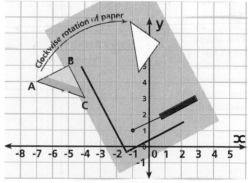

❸ After 90° rotation mark in position of Image

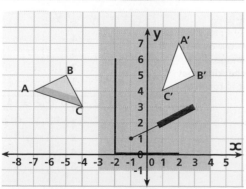

❹ Remove the tracing paper. Draw in image and label

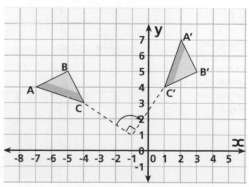

Transformations 3

❸ Translation

A translation alters the position of a shape by moving every point of it by the same distance in the same direction. To describe a translation, you must specify the following two things:

- THE DIRECTION OF THE MOVEMENT
- THE DISTANCE MOVED

This can be summarised using a TRANSLATION VECTOR in which the movement in the x direction is placed directly above the movement in the y direction and both figures are surrounded by brackets. Positive and negative numbers are used to indicate the direction of the movement. (See diagram alongside.) Translation only changes the POSITION of the shape. Everything else stays the same, i.e. length and angle. The object and image are congruent shapes (see page 79).

Difficult translations can be completed more easily if you use tracing paper. Notice that all points (**P'**, **Q'**, **R'**, **S'**, **T'** and **U'**) in the example below have moved 6 to the right and 6 down.

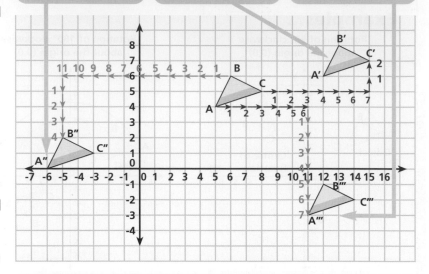

ABC has been translated to A''B''C'' by moving 11 squares to the left and then 4 squares down. This translation is written as a vector $\begin{pmatrix} -11 \\ -4 \end{pmatrix}$

ABC has been translated to A'B'C' by moving 7 squares to the right and then 2 squares up. This translation is written as a vector $\begin{pmatrix} 7 \\ 2 \end{pmatrix}$

ABC has been translated to A'''B'''C''' by moving 6 squares to the right and then 7 squares down. This translation is written as a vector $\begin{pmatrix} 6 \\ -7 \end{pmatrix}$

- **Movement to the right (→) or up (↑) is positive**
- Movement to the left (←) or down (↓) is negative

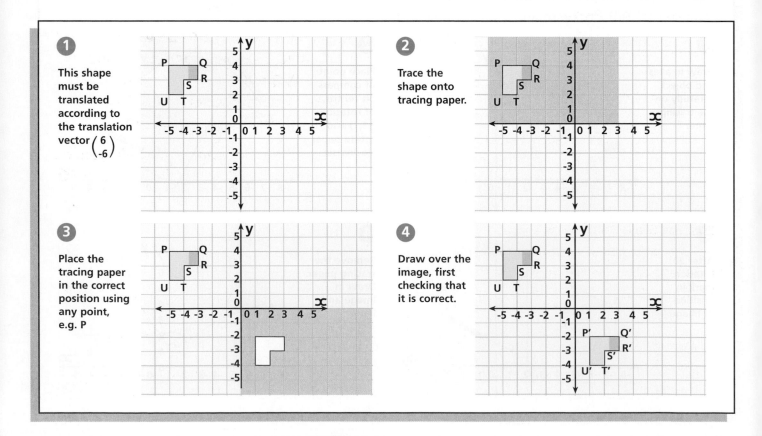

① This shape must be translated according to the translation vector $\begin{pmatrix} 6 \\ -6 \end{pmatrix}$

② Trace the shape onto tracing paper.

③ Place the tracing paper in the correct position using any point, e.g. P

④ Draw over the image, first checking that it is correct.

➍ Enlargement

An enlargement changes the size of a shape. The shape can be made bigger or smaller according to the Scale Factor. All enlargements take place from one point called the Centre of Enlargement.

To describe an enlargement you must specify the following two things:
- THE CENTRE OF ENLARGEMENT
- THE SCALE FACTOR

Positive Scale Factor

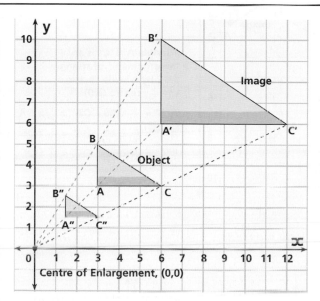

Centre of Enlargement, (0,0)

Positive Scale Factor Greater Than 1

A Scale Factor greater than 1 means that the size of the image is bigger than the size of the object. Triangle A'B'C' is an enlargement of triangle ABC by a scale factor of 2, centre (0,0).

A'B' = 2 x AB		OA' = 2 x OA
A'C' = 2 x AC	and	OB' = 2 x OB
B'C' = 2 x BC		OC' = 2 x OC

Positive Scale Factor Less Than 1

A Scale Factor less than 1 means that the size of the image is smaller than the size of the object. Triangle A''B''C'' is an 'enlargement' of Triangle ABC by a scale factor of $\frac{1}{2}$, centre (0,0).

A''B'' = $\frac{1}{2}$ x AB		OA'' = $\frac{1}{2}$ x OA
A''C'' = $\frac{1}{2}$ x AC	and	OB'' = $\frac{1}{2}$ x OB
B''C'' = $\frac{1}{2}$ x BC		OC'' = $\frac{1}{2}$ x OC

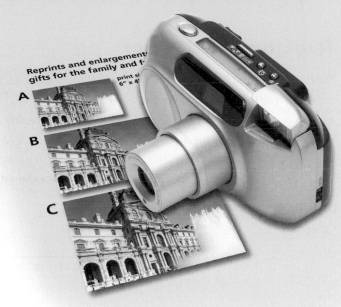

Reprints and enlargements gifts for the family and f...
print si...
6" x 4"...

Enlargement only changes the SIZE of the shape (i.e. the lengths of its sides) and its POSITION. Sometimes you are asked to calculate the Scale Factor and find the Centre of Enlargement.

Example

Triangle P'Q'R' is an enlargement of triangle PQR. What is the Scale Factor of the enlargement and the coordinates of the Centre of Enlargement?

To find the Centre of Enlargement:
Draw dotted lines passing through P and P' (-----), Q and Q' (-----), R and R' (-----). Where these dotted lines cross is the Centre of Enlargement. Coordinates are (0, 1).

To find the Scale Factor of the enlargement:
P'Q' = 10 units, PQ = 2 units ∴ P'Q' = 5 x PQ. Scale Factor of the enlargement is 5.

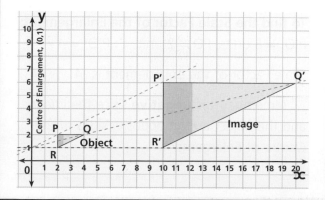

Transformations 5

Negative Scale Factor

With a negative Scale Factor the number part of it tells us if the shape is to be made bigger or smaller, while the negative sign tells us that the image and object are on opposite sides of the Centre of Enlargement.

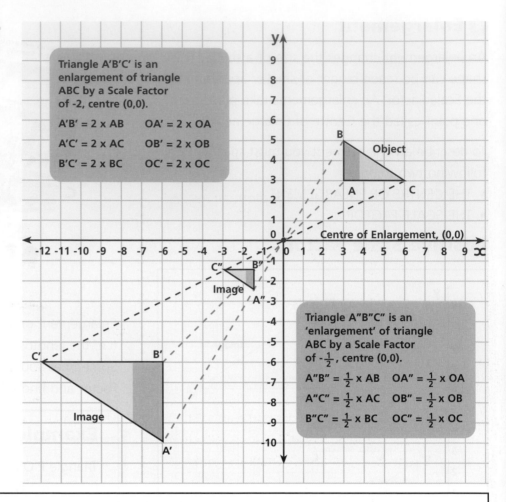

Triangle A'B'C' is an enlargement of triangle ABC by a Scale Factor of -2, centre (0,0).

A'B' = 2 x AB   OA' = 2 x OA

A'C' = 2 x AC   OB' = 2 x OB

B'C' = 2 x BC   OC' = 2 x OC

Centre of Enlargement, (0,0)

Object

Image

Image

Triangle A"B"C" is an 'enlargement' of triangle ABC by a Scale Factor of $-\frac{1}{2}$, centre (0,0).

A"B" = $\frac{1}{2}$ x AB   OA" = $\frac{1}{2}$ x OA

A"C" = $\frac{1}{2}$ x AC   OB" = $\frac{1}{2}$ x OB

B"C" = $\frac{1}{2}$ x BC   OC" = $\frac{1}{2}$ x OC

Example...

Triangle P'Q'R' is an enlargement of triangle PQR. What is the Scale Factor of the enlargement and the coordinates of the Centre of Enlargement?

To find the Centre of Enlargement:
Draw dotted lines passing through P and P' (-----), Q and Q' (-----), R and R' (-----), where these dotted lines cross is the Centre of Enlargement. Coordinates are (1,0)

To find the Scale Factor of the Enlargement:
P'Q' = 8 units, PQ = 2 units ∴ P'Q' = 4 x PQ.

Scale Factor of the enlargement is -4.

... remember it is **negative** since the image and the object are on opposite sides of the Centre of Enlargement.

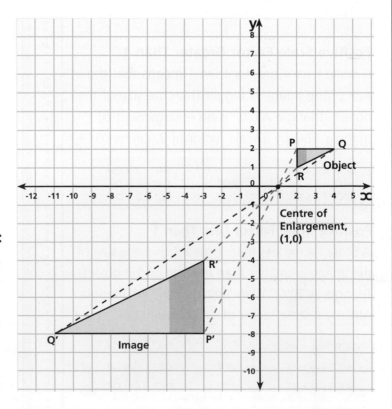

Centre of Enlargement, (1,0)

Object

Image

Transformations 6

Combination of Transformations

Very often a combination of two (or more) transformations can be described by a single transformation.

Examples...

1 Triangle ABC is reflected in the **y**-axis to A'B'C' and then A'B'C' is reflected in the **x**-axis to A''B''C''. Draw the two transformations and describe fully the single transformation that maps triangle ABC onto triangle A''B''C''.

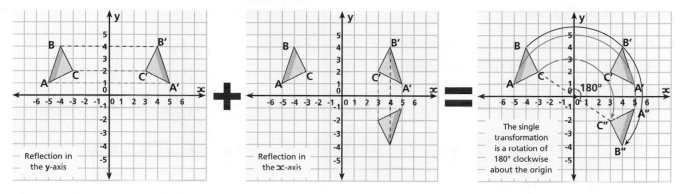

2 Shape A is rotated 90° clockwise about the origin to shape B. Shape B is then reflected in the **x**-axis to shape C. Draw the two transformations and describe fully the single transformation that maps shape A onto shape C.

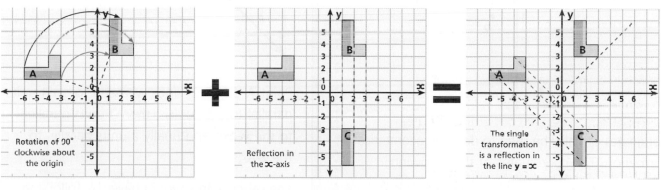

Characteristics of Transformations - A Summary

TRANSFORMATION / CHARACTERISTIC	You need to specify...	Properties that are preserved	Properties that change	Congruent or Similar?
REFLECTION	• The mirror line (equation of the line of reflection)	Shape and Size (e.g. angles, lengths of sides)	Orientation, Position	Congruent
ROTATION	• Direction of turn • Centre of Rotation • Angle turned through	As above	Orientation, Position	Congruent
TRANSLATION	• Direction of movement • Distance moved	As above plus orientation	Position	Congruent
ENLARGEMENT	• Centre of Enlargement • Scale Factor	Angles, ratios of lengths of side, orientation	Position, Size	Similar

Coordinates 1

Points on a Number Line

One coordinate is needed to identify a point on a number line, i.e. in 1-D (in one dimension). On this number line, point A is (-2) and point B is (3).

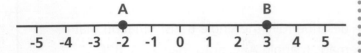

Points in a Plane

Two coordinates are needed to identify a point in a plane, i.e. in 2-D (in two dimensions). On the grid below point A is (3,4), B is (-2,1) C is (-1,-4) and D is (1,-1).

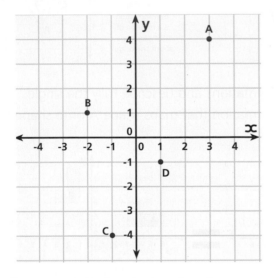

Coordinates can also be used to describe geometrical information. The coordinates of the following points: A is (3,2), B is (2,-2) C is (-3,-2) and D is (-2,2) are the vertices (corners) of a parallelogram.

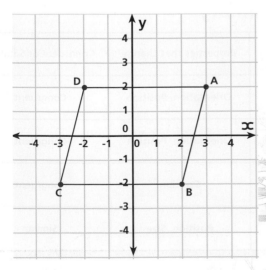

Points in Space

Three coordinates are needed to identify a point in space, i.e. in 3-D (in three dimensions). Three axes each at right angles to each other are needed. The coordinates of each point represent distances from **0**, a fixed point, firstly parallel to the **x**-axis then parallel to the **y**-axis, and finally parallel to the **z**-axis.

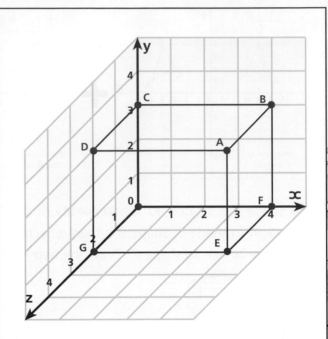

Point A is (4,3,2) since we need to go 4 units parallel to the **x**-axis, then 3 units parallel to the **y**-axis and finally 2 units parallel to the **z**-axis.

The other points would have the following coordinates: B is (4,3,0) C is (0,3,0) D is (0,3,2) E is (4,0,2) F is (4,0,0) and G is (0,0,2)

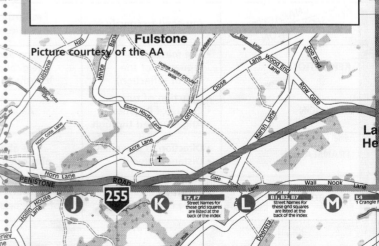

Fulstone
Picture courtesy of the AA

Coordinates 2

Coordinates of the Midpoint of a Line Segment

To find the coordinates of the midpoint, M, of a line segment AB you need to find…

- the average of the x coordinates of points A and B. This is the x coordinate of the midpoint, M.
- the average of the **y** coordinates of points A and B. This is the **y** coordinate of the midpoint, M.

Examples…

 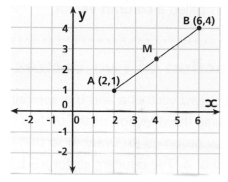

Average of x coordinates of points A and B

$$= \frac{2+6}{2} = \frac{8}{2} = 4$$

Average of **y** coordinates of points A and B

$$= \frac{1+4}{2} = \frac{5}{2} = 2.5$$

Coordinates of point M are (4,2.5)

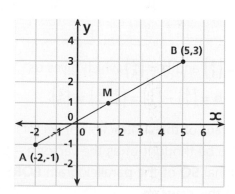

Average of x coordinates of points A and B

$$= \frac{-2+5}{2} = \frac{3}{2} = 1.5$$

Average of **y** coordinates of points A and B

$$= \frac{-1+3}{2} = \frac{2}{2} = 1$$

Coordinates of point M are (1.5,1)

Calculating the Length of a Line Segment

The length of a line segment AB can be found using Pythagoras' Theorem. Firstly complete the right-angled triangle ABC where AB represents the hypotenuse. You then find the length of AC and BC using the scale on the axes (not by using a ruler).

Examples…

 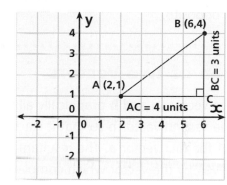

Using Pythagoras' Theorem

$$AB^2 = AC^2 + BC^2$$
$$= 4^2 + 3^2$$
$$= 16 + 9$$
$$= 25$$
$$AB = \sqrt{25}$$
$$= 5 \text{ units}$$

 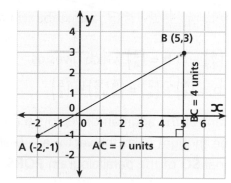

Using Pythagoras' Theorem

$$AB^2 = AC^2 + BC^2$$
$$= 7^2 + 4^2$$
$$= 49 + 16$$
$$= 65$$
$$AB = \sqrt{65} \text{ units (in surd form)}$$
$$= 8.1 \text{ units (to 2 sig. fig.)}$$

Vectors 1

Vectors are quantities that have both **magnitude** (or **size**) and **direction**.

A common example of a vector quantity is displacement, which is distance moved in a particular direction. Another example is velocity, which is speed in a particular direction.

A vector is usually drawn as a **line with an arrow on it** where the length of the line represents magnitude (size) and the arrow represents direction. **A** and **B** are two points as shown opposite.

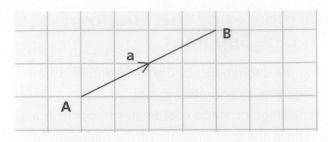

The displacement of **A** to **B** can be described in the following way:

$$\overrightarrow{AB} = a = \begin{pmatrix} 4 \\ 2 \end{pmatrix}$$

To get from A to B we have to go 4 across and then 2 up. $\begin{pmatrix} 4 \\ 2 \end{pmatrix}$ is called a **column vector**.

Addition of Two Vectors

To add two vectors **a** and **b** graphically all we do is draw the second vector **b** so that it starts at the end of the first vector **a**. The resultant vector **a + b** is given by the vector which completes the triangle set up by **a** and **b**.

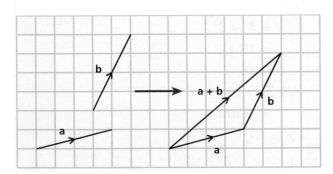

Also, $\mathbf{a} = \begin{pmatrix} 4 \\ 1 \end{pmatrix}$ and $\mathbf{b} = \begin{pmatrix} 2 \\ 4 \end{pmatrix}$ which gives us:

$$\mathbf{a} + \mathbf{b} = \begin{pmatrix} 4 \\ 1 \end{pmatrix} + \begin{pmatrix} 2 \\ 4 \end{pmatrix} = \begin{pmatrix} 4+2 \\ 1+4 \end{pmatrix} = \begin{pmatrix} 6 \\ 5 \end{pmatrix}$$

The resultant of the addition of vectors **a** and **b** can also be thought of as the vector **a + b** along the diagonal **SQ** of the parallelogram **PQRS** as shown in this diagram.

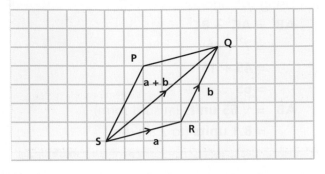

Subtraction of Two Vectors

To subtract two vectors **a** and **b** graphically you firstly reverse the direction of the second vector so that **b** now becomes **-b**. Draw **-b** so that it starts at the end of **a**. The resultant vector **a – b** is again given by the vector which completes the triangle set up by **a** and **-b**.

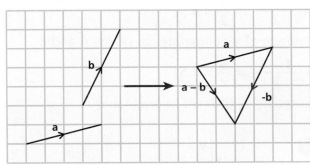

Also, $\mathbf{a} = \begin{pmatrix} 4 \\ 1 \end{pmatrix}$ and $\mathbf{b} = \begin{pmatrix} 2 \\ 4 \end{pmatrix}$ which gives us:

$$\mathbf{a} - \mathbf{b} = \begin{pmatrix} 4 \\ 1 \end{pmatrix} - \begin{pmatrix} 2 \\ 4 \end{pmatrix} = \begin{pmatrix} 4-2 \\ 1-4 \end{pmatrix} = \begin{pmatrix} 2 \\ -3 \end{pmatrix}$$

The resultant of the subtraction of vectors **a** and **b** can also be thought of as the vector **a – b** along the diagonal **SQ** of the parallelogram **PQRS** as shown in this diagram.

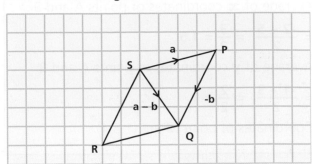

Vectors 2

Relationship Between Two Vectors

Two vectors are equal if they have the **same magnitude (size)** and **direction**. It is also possible for two vectors to have **different magnitudes** and the **same direction**.

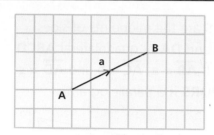

If we take: $\overrightarrow{AB}$ = a = $\binom{4}{2}$
We will now compare the magnitude and direction of $\overrightarrow{AB}$ to four other vectors.

1

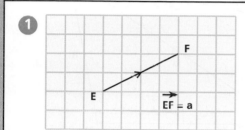

$\overrightarrow{EF}$ = $\binom{4}{2}$ = $\overrightarrow{AB}$ = a. Therefore $\overrightarrow{EF}$ and $\overrightarrow{AB}$...
i) Have the **same magnitude**, i.e. **EF** and **AB** are the same length.
ii) Are in the **same direction**, i.e. **EF** and **AB** are parallel lines.

2

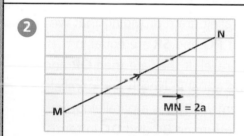

$\overrightarrow{MN}$ = $\binom{8}{4}$ = 2 x $\binom{4}{2}$ = 2$\overrightarrow{AB}$ = 2a. Therefore $\overrightarrow{MN}$ and $\overrightarrow{AB}$...
i) Have **different magnitudes** with **MN** twice the length of **AB**.
ii) Are in the **same direction**, i.e. **MN** and **AB** are parallel lines.

3

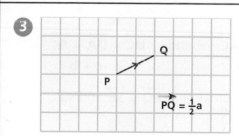

$\overrightarrow{PQ}$ = $\binom{2}{1}$ = $\frac{1}{2}$ x $\binom{4}{2}$ = $\frac{1}{2}$ $\overrightarrow{AB}$ = $\frac{1}{2}$a. Therefore $\overrightarrow{PQ}$ and $\overrightarrow{AB}$...
i) Have **different magnitudes** with **PQ** half the length of **AB**.
ii) Are in the **same direction**, i.e. **PQ** and **AB** are parallel lines.

4

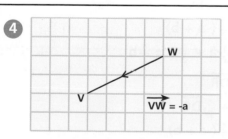

$\overrightarrow{VW}$ = $\binom{-4}{-2}$ = -1 x $\binom{4}{2}$ = -$\overrightarrow{AB}$ = -a. Therefore $\overrightarrow{VW}$ and $\overrightarrow{AB}$...
i) Have the **same magnitude**, i.e. **VW** and **AB** are the same length.
ii) Are in **opposite directions** (the - sign indicates opposite directions). This still means that **VW** and **AB** are parallel lines.

In Summary...

If 'one vector' is equal to a fraction or multiple of 'another vector' then...
1. Their lengths are related by a factor equal to the fraction or multiple (which is 1 if they are the same length).
2. They are parallel.

Vector Geometry

As we have seen, the displacement between any two points can be described by a vector. The displacement between another two points can also be described by a vector and comparisons can be made between these two vectors in terms of their magnitude and direction.

Examples...

1 In triangle **ABC**, **G** is the midpoint of **AB** and **H** is the midpoint of **AC**. If $\overrightarrow{AB}$ = **a** and $\overrightarrow{AC}$ = **b**...

a) Find expressions for $\overrightarrow{BC}$ and $\overrightarrow{GH}$ in terms of **a** and **b**.

b) Using your answers to part **a)** what can you prove about **BC** and **GH**?

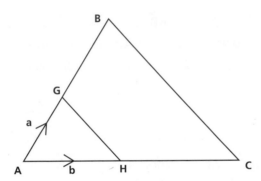

a) $\overrightarrow{BC} = \overrightarrow{BA} + \overrightarrow{AC}$
    $= -a + b$

> To get from **B** to **C** we have to go from **B** to **A** ($\overrightarrow{BA}$) and then from **A** to **C** ($\overrightarrow{AC}$). Also $\overrightarrow{BA} = -\overrightarrow{AB} = -a$

$\overrightarrow{GH} = \overrightarrow{GA} + \overrightarrow{AH}$
    $= -\frac{1}{2}a + \frac{1}{2}b$

> To get from **G** to **H** we have to go from **G** to **A** ($\overrightarrow{GA}$) and then from **A** to **H** ($\overrightarrow{AH}$). Also $\overrightarrow{GA} = -\overrightarrow{AG} = -\frac{1}{2}\overrightarrow{AB} = -\frac{1}{2}a$ and $\overrightarrow{AH} = \frac{1}{2}\overrightarrow{AC} = \frac{1}{2}b$

b) Since...

BC = -a + b

and

GH = $-\frac{1}{2}a + \frac{1}{2}b$
    $= \frac{1}{2}(-a + b)$... $\frac{1}{2}$ is a common factor

This now gives us...
$\overrightarrow{GH} = \frac{1}{2}\overrightarrow{BC}$ ... since $\overrightarrow{BC}$ = -a + b

And so this proves...
i) **GH** = $\frac{1}{2}$**BC** i.e. **GH** is half the length of **BC**
ii) **GH** is parallel to **BC**

2 P, Q, R and S are the midpoints of sides **AB**, **BC**, **CD** and **DA** of quadrilateral **ABCD** respectively. If $\overrightarrow{AB}$ = a, $\overrightarrow{AC}$ = b and $\overrightarrow{AD}$ = c prove that **PQRS** is a parallelogram.

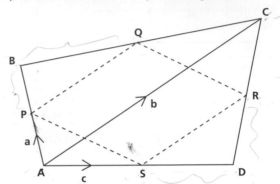

PQRS is a parallelogram providing that we can prove that **PQ** and **SR** are equal in length and parallel. If they are then **QR** and **PS** will also be equal in length and parallel and consequently **PQRS** is a parallelogram. Firstly let us determine $\overrightarrow{BC}$ and $\overrightarrow{DC}$ in terms of **a**, **b** and **c**.

$\overrightarrow{BC} = \overrightarrow{BA} + \overrightarrow{AC}$ and $\overrightarrow{DC} = \overrightarrow{DA} + \overrightarrow{AC}$
    $= -a + b$ $= -c + b$

We will now prove that $\overrightarrow{PQ} = \overrightarrow{SR}$ and so...
$\overrightarrow{PQ} = \overrightarrow{PB} + \overrightarrow{BQ}$
    $= \frac{1}{2}\overrightarrow{AB} + \frac{1}{2}\overrightarrow{BC}$
    $= \frac{1}{2}a + \frac{1}{2}(-a + b)$
    $= \frac{1}{2}a - \frac{1}{2}a + \frac{1}{2}b = \frac{1}{2}b$

$\overrightarrow{SR} = \overrightarrow{SD} + \overrightarrow{DR}$
    $= \frac{1}{2}\overrightarrow{AD} + \frac{1}{2}\overrightarrow{DC}$
    $= \frac{1}{2}c + \frac{1}{2}(-c + b)$
    $= \frac{1}{2}c - \frac{1}{2}c + \frac{1}{2}b = \frac{1}{2}b$

$\overrightarrow{PQ} = \overrightarrow{SR}$, which means that **PQ** and **SR** are equal in length and are parallel; **PQRS** is therefore a parallelogram.

Perimeter

This is a measure of the distance all the way around the outside of a shape. All we need to know are the lengths of all the sides that make up the shape and then add them together.

Examples...

 1

The lengths of all the sides are given or can be worked out and so...

Perimeter

= Length of AB + BC + CD + DE + EF + FA

= 6cm + 2.5cm + 4cm + (4.1 − 2.5)cm + 2cm + 4.1cm

= **20.2cm** (remember the units)

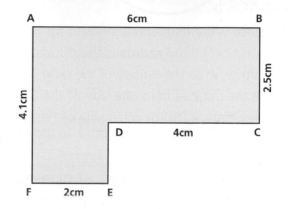

 2

The lengths of all the sides have to be measured using a ruler and so...

Perimeter

= Length of AB + BC + CA

= 6.2cm + 5.7cm + 4.3cm

= **16.2cm** (remember the units)

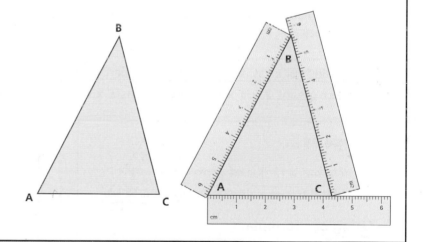

 3

The lengths of all the sides can be found by counting squares and so...

Perimeter

= Length of AB + BC + CD + DE + EF + FA

= 1cm + 3cm + 1cm + 1cm + 2cm + 4cm

= **12cm** (remember the units)

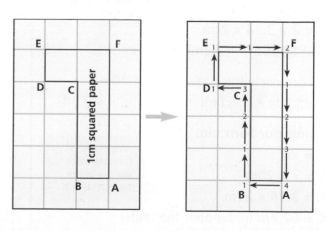

Circumference of a Circle

Circumference

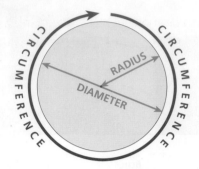

Circumference is the mathematical word for the perimeter of a circle. The **radius** is the distance from the centre, •, to the outside of the circle. The **diameter** is the distance from one side of the circle through the centre, •, to the other side of the circle, which therefore means that…

> **Length of Radius = $\frac{1}{2}$ x Length of Diameter**
> or **Length of Diameter = 2 x Length of Radius**

The circumference can be found by using this formula:

> **Circumference = 2πr** or **Circumference = πd**

Where π (called 'pi') has an approximate value of 3.14 and **2πr** means **2 x π x r** and **πd** means **π x d**.

Examples…

1 A circle has a radius of 2cm.
Calculate its circumference using π ≈ 3.14

Using our formula:

C = 2πr

C = 2 x π x 2cm

> We use this formula since we are given the RADIUS

C = 2 x 3.14 x 2cm

C = 12.56cm (remember the units)

2 A corn circle has a diameter of 20m.
Calculate its circumference using π ≈ 3.14

Using our formula:

C = πd

C = π x 20m

> We use this formula since we are given the DIAMETER

C = 3.14 x 20m

C = 62.8m (remember the units)

Length of an Arc

An arc is simply part of the circumference of a circle.

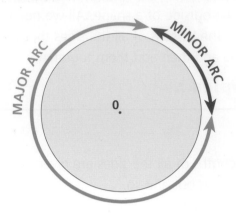

The length of an arc is given by:

> **Length of Arc = $\frac{\theta}{360°}$ x 2πr**

Where θ is the angle at the centre of the circle.

Example…

Calculate the length of the minor arc **AB** in the following diagram using π ≈ 3.14

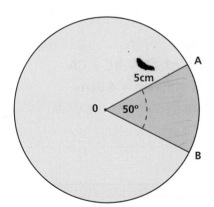

Length of minor arc

$= \frac{50°}{360°}$ x 2πr

> This is the fraction of the circumference of the circle taken up by the arc.

$= \frac{50°}{360°}$ x 2 x 3.14 x 5

= **4.36cm** (to 2 d.p.)

> **Note:** The length of the major arc would be:
> $= \frac{310°}{360°}$ x 2πr

Area 1

Area

This is a measure of the amount of surface a two-dimensional shape covers. Area is usually measured in **units²**, e.g. **cm²** (cm squared) or **m²** (m squared) etc.

Estimation of Area Using Squared Paper

The area of a shape can be estimated if it has been drawn on squared paper. All we have to do is count the number of squares taken up by the shape.

This method is particularly useful when we have irregular shapes, although our answer will be an estimate of the area and not an exact value.

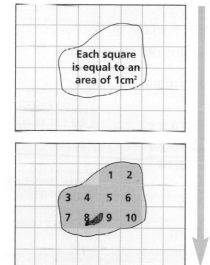

Each square is equal to an area of 1cm²

Count the whole squares and any squares more than half covered

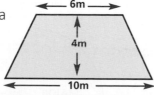

Area = 10cm²

Areas of Common Shapes

The following shapes each have a formula which can be used to work out their area exactly.

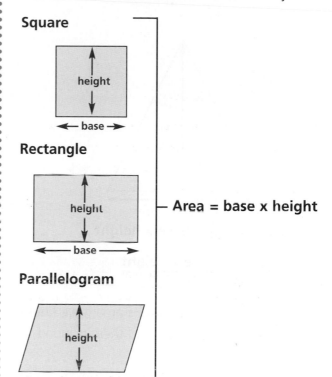

Square

height

base

Rectangle

height

base

Parallelogram

height

base

Area = base x height

Trapezium

b

height (h)

a

Area = $\frac{1}{2}$ (a + b) h

or... = $\frac{(a + b)}{2}$ x h

Examples...

1 Calculate the area of the following trapezium.

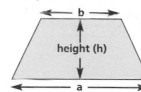

6m

4m

10m

Using our formula: Area = $\frac{(a + b)}{2}$ x h

Area = $\frac{(10m + 6m)}{2}$ x 4m

= $\frac{16m}{2}$ x 4m

= 8m x 4m

= **32m²** (remember the units)

2 The following parallelogram has an area of 24cm². Calculate its height if the length of its base is 10cm.

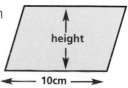

height

10cm

Using our formula: Area = Base x Height

24cm² = 10cm x Height

$\frac{24}{10}$ = $\frac{\cancel{10}}{\cancel{10}}$ x Height

Divide both sides by 10 to give us height on its own

Height = **2.4cm** (remember the units)

Area 2

Area of a Triangle

The area of a triangle can be found very simply, if we know the length of its base and height, using the following formula:

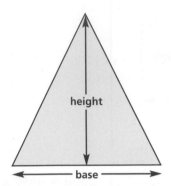

$$\text{Area} = \frac{1}{2} \times \text{base} \times \text{height}$$

$$\text{or...} = \frac{\text{base} \times \text{height}}{2}$$

However, we can also work out its area if we know the length of two of the sides and their included angle. The formula is:

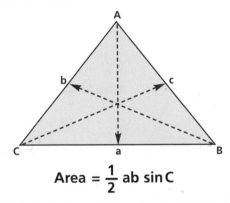

$$\text{Area} = \frac{1}{2} \, ab \sin C$$

Example...

Calculate the area of the following triangle:

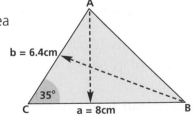

$\text{Area} = \frac{1}{2} \, ab \sin C$

$= \frac{1}{2} \times 8\text{cm} \times 6.4\text{cm} \times \sin 35°$

$= \frac{1}{2} \times 8 \times 6.4 \times 0.5736$

$= 14.68\text{cm}^2$

Area of a Circle

The area of a circle is given by the formula:

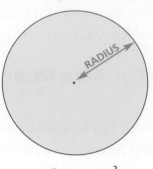

$$\text{Area} = \pi r^2$$

Yet again π has an approximate value of **3.14** and πr^2 means π **x radius squared** or π **x r x r**.

Example...

A circular cricket field has an area of 20 096m². Calculate its radius using $\pi \approx 3.14$.

Using our formula:

$\text{Area} = \pi r^2$

$20\,096\text{m}^2 = 3.14 \times r^2$

$\dfrac{20\,096\text{m}^2}{3.14} = \dfrac{\cancel{3.14} \times r^2}{\cancel{3.14}}$

> Divide both sides by 3.14 to leave r^2 on its own

$6\,400\text{m}^2 = r^2$

$\sqrt{6\,400\text{m}^2} = \sqrt{r^2}$

> Take the square root of both sides to leave r on its own

$r = 80\text{m}$

Area of a Sector

A sector is the area enclosed by two radii and an arc. It can be major or minor.

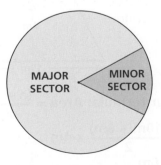

The area of a sector is given by:

$$\text{Area of Sector} = \frac{\theta}{360°} \times \pi r^2$$

where θ is the angle at the centre of the circle.

Example...

Calculate the area of the minor sector in the following diagram using $\pi \approx 3.14$

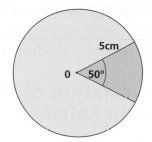

Area of Minor Sector $= \frac{50°}{360°} \times \pi r^2$

$= \frac{50°}{360°} \times 3.14 \times 5^2$

$= 10.90cm^2$ (to 2 d.p.)

NOTE: the area of the major sector would be $\frac{310°}{360°} \times \pi r^2$

Area of a Segment

Segments are formed by a chord (see page 121). They can be major or minor.

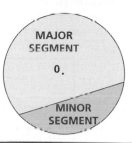

Example...

Calculate the area of the minor segment in the following diagram using $\pi \approx 3.14$

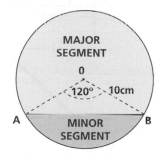

We can work it out in two stages:

1 Area of Sector **ABO** $= \frac{120°}{360°} \times \pi \times 10^2$

$= 104.67cm^2$ (to 2 d.p.)

2 Area of Triangle **ABO** $= \frac{1}{2}$ **ab sin C**

$= \frac{1}{2} \times 10 \times 10 \times \sin 120°$

$= 43.30cm^2$ (to 2 d.p.)

Area of Segment = Area of **1** − Area of **2**

$= 104.67cm^2 - 43.30cm^2$

$= 61.37cm^2$

Surface Area of Solids

The surface area of any solid is simply the area of the NET which can be folded to completely cover the outside of the solid (see page 111).

Example...

1 Calculate the surface area of the following triangular prism which has an equilateral triangle as its cross-section.

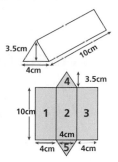

Area of 1 (Rectangle) $= 10cm \times 4cm = 40cm^2$
Therefore Area of 2 + 3 $= 80cm^2$

Area of 4 (Triangle) $= \frac{4cm \times 3.5cm}{2} = 7cm^2$

Therefore Area of 5 $= 7cm^2$

Surface Area of prism
= Area of 1 + 2 + 3 + 4 + 5
$= 40cm^2 + 80cm^2 + 7cm^2 + 7cm^2$
$= 134cm^2$ (remember the units)

2 Calculate the surface area of this cylinder using $\pi \approx 3.14$

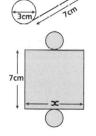

Area of each end $= \pi r^2$
$= 3.14 \times 1.5^2$
$= 7.065cm^2$

Therefore both ends
$= 7.065 \times 2 = 14.13cm^2$

Length of x = circumference of circle
$= 2\pi r$
$= 2 \times 3.14 \times 1.5$
$= 9.42cm$

Therefore area of rectangle
$= 7 \times 9.42$
$= 65.94cm^2$

Therefore total area
$= 14.13cm^2 + 65.94cm^2$
$= 80.07cm^2$

Area 4 / Volume 1

Surface Area of a Sphere

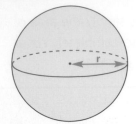

Surface Area = $4\pi r^2$

Curved Surface Area of a Cone

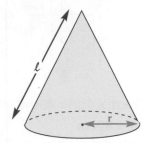

Curved Surface Area = $\pi r \ell$

(N.B. The total surface of the cone is $\pi r \ell + \pi r^2$ where πr^2 is the area of the base).

Surface Area of a Pyramid

There is no general formula for the surface area. Draw its net (see page 111) and calculate each area separately.

Areas of Compound Shapes

These are shapes that can be divided up into smaller shapes. The area of each of these smaller shapes can then be calculated and added together.

Example...

The diagram shows the layout for a side wall of a house. Calculate its area.

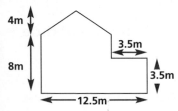
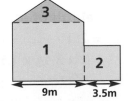

Area of 1 (Rectangle) = 9m x 8m = 72m²

Area of 2 (Square) = 3.5m x 3.5m = 12.25m²

Area of 3 (Triangle) = $\dfrac{9m \times 4m}{2}$ = 18m²

Area of wall = Area 1 + Area 2 + Area 3

= 72m² + 12.25m² + 18m²

= **102.25m²** (remember the units)

Volume

Volume is a measure of the amount of space a 3-D object takes up. It is usually measured in **units³**, e.g. **cm³** (cm cubed) or **m³** (m cubed).

Calculation of the Volume of a Solid Made Up of Cubes

Providing we know the volume of one cube then all we have to do is work out how many cubes there are in each layer of the solid and then add them up.

Example

In the following example each cube has a volume of 1cm³.

Solid is made up of TWO LAYERS

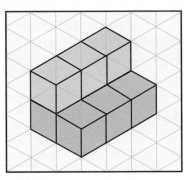

Bottom layer

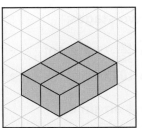

+

Top layer

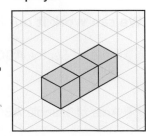

Volume = 6 cubes + 3 cubes

= 6cm³ + 3cm³

= 9cm³

Volume 2

Volume of a Cuboid

> **Volume = length x width x height**
> **V = ℓ x w x h**

Example

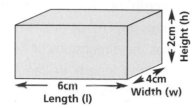

Using our formula:

Volume = length x width x height

$\quad$ = 6cm x 4cm x 2cm

$\quad$ = 48cm^3 (remember the units)

Volume of a Prism

A prism is a solid which has a uniform cross-section from one end of the solid to the other end.

> **Volume of a prism**
> **= area of cross-section x length**

Example

Calculate the volume of the triangular prism.

The cross-section is obviously a TRIANGLE

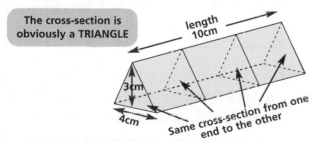

length 10cm

3cm

4cm

Same cross-section from one end to the other

Using our formula:

Volume = area of cross-section x length

$\quad = \dfrac{\text{(base x height)}}{2}$ x length

$\quad = \dfrac{\text{(4cm x 3cm)}}{2}$ x 10cm

$\quad$ = 6cm^2 x 10cm

$\quad$ = 60cm^3 (remember the units)

Volume of a Cylinder

A cylinder is a prism which has a uniform cross-section of a circle from one end of the prism to the other. The volume of any cylinder is given by the formula:

> **Volume of a cylinder = $\pi r^2\ell$ = πr^2 x ℓ**
> (where ℓ is the length of the cylinder)

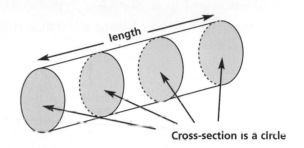

length

Cross-section is a circle

Examples...

1 Calculate the volume of the following cylinder. (let π = 3.14)

8cm

2cm

Using our formula:

Volume = $\pi r^2\ell$

Volume = πr^2 x length

$\quad$ = 3.14 x (2cm)^2 x 8cm

$\quad$ = 3.14 x 4cm^2 x 8cm

$\quad$ = 100.48 cm^3 (remember the units)

2 A cylindrical tank is 1.6m long and holds 0.8m^3 of oil when full. What is the radius of the cylinder (let π = 3.14)?

1.6m

Using our formula:

Volume = $\pi r^2\ell$

$\quad$ 0.8m^3 = 3.14 x r^2 x 1.6m

$\quad$ 0.8m^3 = 5.024m x r^2

$\quad \dfrac{0.8m^3}{5.024m} = \dfrac{5.024m}{5.024m}$ x r^2

Divide both sides by 5.024 to leave r^2 on its own

$\quad$ 0.159m^2 = r^2

$\quad \sqrt{0.159m^2} = \sqrt{r^2}$

Take square root of both sides to leave r on its own

$\quad\quad$ r = 0.40m (to 2 d.p.)

(remember the units)

Volume 3

Volume of a Sphere

$$\text{Volume} = \frac{4}{3} \times \pi \times (\text{radius})^3$$
$$= \frac{4}{3} \pi r^3$$

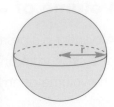

Volume of a Pyramid

$$\text{Volume} = \frac{1}{3} \times \text{Area of Base} \times \text{Vertical Height}$$

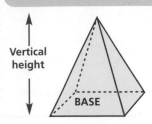 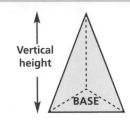

Vertical height — BASE

Vertical height — BASE

Volume of a Cone

$$\text{Volume} = \frac{1}{3} \times \text{Area of Base} \times \text{Vertical Height}$$
$$= \frac{1}{3} \pi r^2 h$$

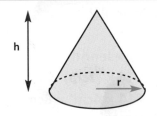

Example...

Calculate the volume of the solid opposite.

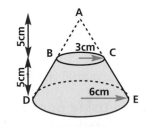

Volume of solid is the Volume of Cone **ADE** – Volume of Cone **ABC**.

Volume of Cone **ADE** $= \frac{1}{3} \times \pi \times 6^2 \times 10$
$= 376.8\text{cm}^3$

Volume of Cone **ABC** $= \frac{1}{3} \times \pi \times 3^2 \times 5$
$= 47.1\text{cm}^3$

Volume of Solid $= 376.8\text{cm}^3 - 47.1\text{cm}^3$
$= 329.7\text{cm}^3$

Distinguishing between Formulae for Perimeter, Area and Volume

Any formula or expression for…

Perimeter - has only a single dimension 'length' in it, since its units are cm or m.

Area - has the dimensions of 'length x length' or 'length^2' in it, since its units are cm^2 or m^2.

Volume - has the dimensions of 'length x length x length' or 'length^3' in it, since its units are cm^3 or m^3.

All numbers and symbols (e.g. π) have no dimensions and therefore don't contribute to the units.

Important Point!

In your exam you may well be asked to distinguish between expressions representing length or perimeter, area, volume. These questions may take the following form:

Examples...

1 In the expressions in the table below, **a**, **b**, **c** and **d** represent lengths. All numbers and symbols have no dimensions. Tick the boxes beneath the expressions which could represent areas.

$\dfrac{\pi dcb}{2a}$	$\dfrac{a^3}{2}$	$4a^2$	$a^3 + b$	$\dfrac{a + b}{c}$	$3a^2 + b^2$	$3a^2b^2$
✓		✓				✓

2 In the expressions in the table below, **p**, **q** and **r** represent lengths. Tick the appropriate box to indicate whether the expression represents a length, an area, a volume or none of these.

Expression	Length	Area	Volume	None
pq + qr		✓		
p x q x r			✓	
p + q + r	✓			
p^{2}qr				✓

Similar figures are identical in their shape but they are not identical in their size (they can be bigger or smaller). The two figures (cuboids) below are similar.

Figure A

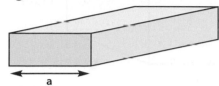

Figure B

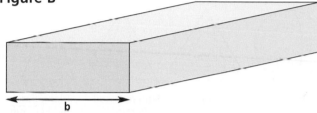

From our two figures:

The ratio of two corresponding lengths is **a : b** (all other corresponding lengths will also be in the same ratio).

Since **area** is a measure of 'length squared', e.g. cm^2 or m^2, then the ratio of their areas is the ratio of the two corresponding lengths squared.

To put it simply...
Ratio of their areas = $a^2 : b^2$

Also, since **volume** is a measure of 'length cubed', e.g. cm^3 or m^3, then the ratio of their volumes is the ratio of the two corresponding lengths cubed.

To put it simply...
Ratio of their volumes = $a^3 : b^3$

Example...

In Figure A, a = 4cm and in Figure B, b = 6cm. If the surface area of Figure A is $52cm^2$ and its volume is $24cm^3$ calculate the surface area, and the volume of Figure B.

We know that...
Ratio of two corresponding lengths, **a : b**, is **4cm : 6cm** which is the equal to **2 : 3**.

① Ratio of areas = $a^2 : b^2$
  = $2^2 : 3^2$
  = **4 : 9** (i.e. 4 parts to 9 parts)

  For Figure **A**, 4 parts = $52cm^2$
  1 part = $\frac{52}{4}$ = $13cm^2$

  Figure **B** is equal to 9 parts
  = $9 \times 13cm^2$
  = $117cm^2$

  ∴ **Surface area of Figure B = $117cm^2$**

② Ratio of volumes = $a^3 : b^3$
  = $2^3 : 3^3$
  = **8 : 27**

  For Figure **A**, 8 parts = $24cm^3$
  1 part = $\frac{24}{8}$ = $3cm^3$

  Figure **B** is equal to 27 parts
  = $27 \times 3cm^3$
  = $81cm^3$

  ∴ **Volume of Figure B = $81cm^3$**

Two important notes:

① If the ratio of the area of two similar figures is for example **9 : 16 ($a^2 : b^2$)** then the ratio of corresponding lengths will be $\sqrt{9} : \sqrt{16}$ = **3 : 4**. This means that to go from area to length you must square root the ratio for area.

The ratio of their volumes will now be
$3^3 : 4^3$ = **27 : 64 ($a^3 : b^3$)**

② If the ratio of the volume of two similar figures is for example **8 : 125 ($a^3 : b^3$)** then the ratio of corresponding lengths will be $\sqrt[3]{8} : \sqrt[3]{125}$ = **2 : 5**. This means that to go from volume to length you must cube root the ratio for volume.

The ratio of their areas will now be
$2^2 : 5^2$ = **4 : 25 ($a^2 : b^2$)**

3-D Shapes 1

Solids

A solid is a three-dimensional (3-D) shape. A very simple solid is the CUBE (a box with all its sides equal in length).

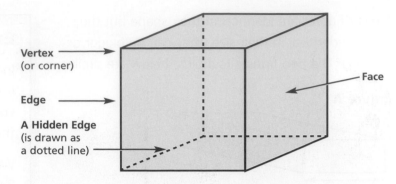

Vertex (or corner)

Edge

A Hidden Edge (is drawn as a dotted line)

Face

Here are some more examples of common solids:

Types of Solid

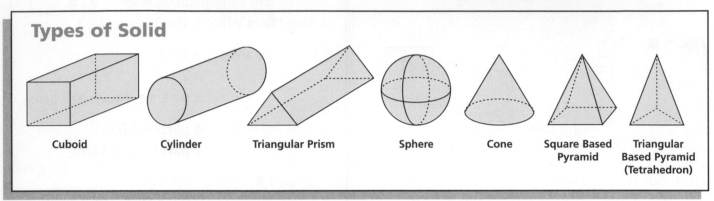

| Cuboid | Cylinder | Triangular Prism | Sphere | Cone | Square Based Pyramid | Triangular Based Pyramid (Tetrahedron) |

Drawing Solids Using Isometric Paper

A disadvantage of drawing solids like the ones above is that accurate measurements of all sides cannot be taken from the diagram. One way of drawing solids is to use isometric paper. This is a grid of equilateral triangles or dots. All solids can be drawn accurately and all measurements can be taken from the diagram.

Examples...

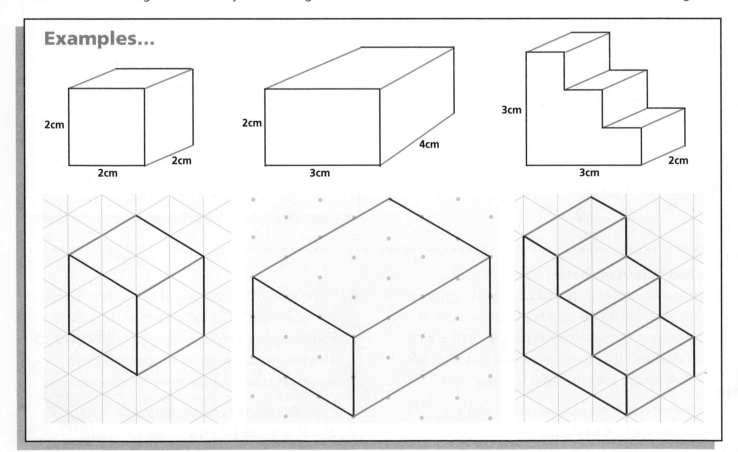

2cm, 2cm, 2cm

2cm, 3cm, 4cm

3cm, 3cm, 2cm

Plans and Elevations

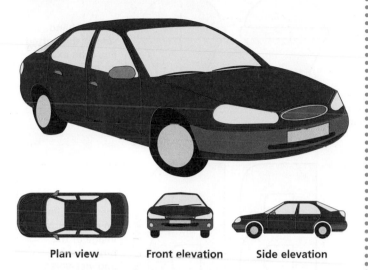

Plan view   Front elevation   Side elevation

This is a picture of a car (a three-dimensional view of a solid). It is possible for us to have THREE different views of the car:

- PLAN VIEW where we look down on the car from above
- FRONT ELEVATION where we look at the car from the front
- SIDE ELEVATION where we look at the car from the side

Example

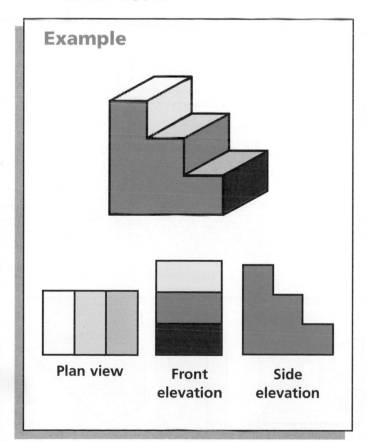

Plan view   Front elevation   Side elevation

Nets for Solids

A NET is a two-dimensional shape which can be folded to completely cover the outside of a solid.

Example

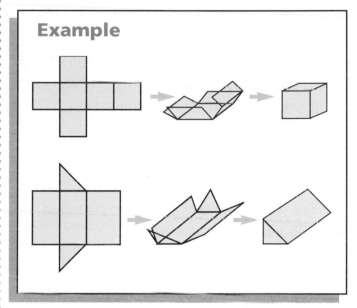

Plane Symmetry

A solid has plane symmetry if it can be 'cut in half' so that one half of the solid is an exact mirror image of the other half.

Examples...

A cuboid has three planes of symmetry

A square based pyramid has four planes of symmetry

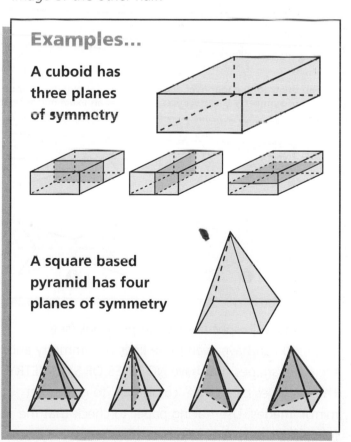

Symmetry 1

Line or Reflective Symmetry

A two-dimensional shape has a LINE OF SYMMETRY if it can be 'cut in half' so that one half of the shape is an exact mirror image of the other half of the shape.

The shapes opposite have 1 line of symmetry. These shapes can be 'cut in half' only once, and one half of the shape is congruent to the other half.

It is also possible for shapes to have more than 1 line of symmetry. The shapes below can be 'cut in half' more than once.

A simple way to find lines of symmetry is to use tracing paper.

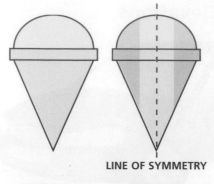

The left hand side of the dotted line is an exact mirror image of the right hand side and vice-versa

LINE OF SYMMETRY

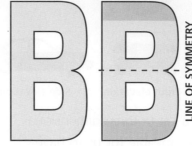

The top side of the dotted line is an exact mirror image of the bottom side and vice-versa

---

Two Lines of Symmetry

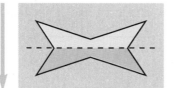

Draw in where you think the line of symmetry is and trace one side of your shape carefully on the tracing paper.

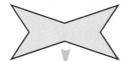

Flip the tracing paper over about the line of symmetry. If your line of symmetry is correct you should get an exact mirror image.

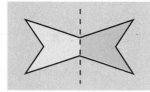

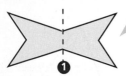 ❶   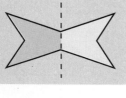 ❷

Three Lines of Symmetry

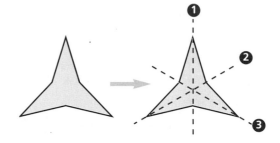

Four Lines of Symmetry

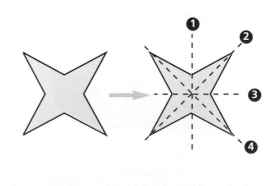

Some shapes, especially regular polygons (see page 78) can have even more lines of symmetry and there are shapes that have NO LINES OF SYMMETRY. These shapes cannot be 'cut in half' to give exact mirror images. Use tracing paper to check that the following shapes have no lines of symmetry.

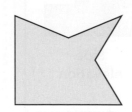  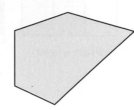

Rotational Symmetry

A two-dimensional shape has ROTATIONAL SYMMETRY if it can be 'rotated about a point', called the Centre of Rotation, to a different position so that it looks the same as it was to begin with. The Order of Rotational Symmetry is equal to the number of times a shape fits onto itself in one 360° turn.

The following shape has ROTATIONAL SYMMETRY ORDER 2. This shape looks the same as its original position every time it is rotated half a turn (180°).

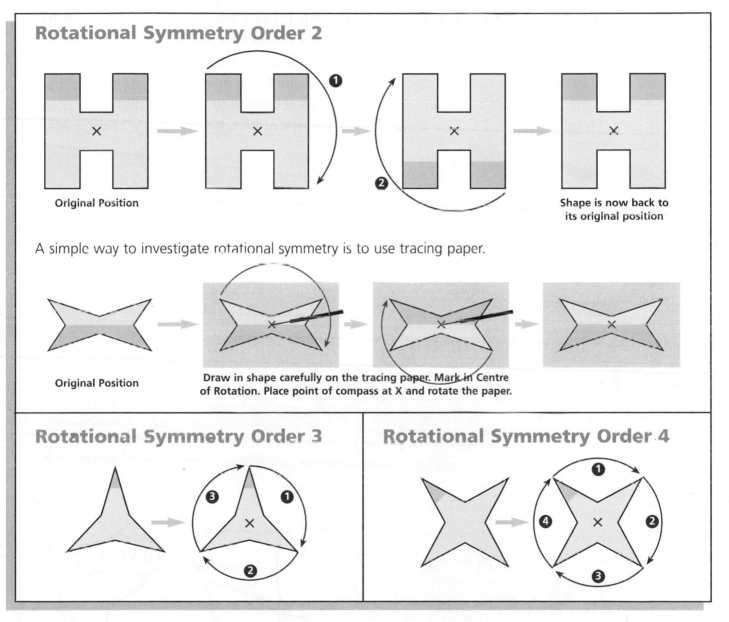

Rotational Symmetry Order 2

Original Position

Shape is now back to its original position

A simple way to investigate rotational symmetry is to use tracing paper.

Original Position

Draw in shape carefully on the tracing paper. Mark in Centre of Rotation. Place point of compass at X and rotate the paper.

Rotational Symmetry Order 3

Rotational Symmetry Order 4

Some shapes, especially regular polygons (see p.78) can have rotational symmetry of even higher orders and there are shapes that have ROTATIONAL SYMMETRY ORDER 1 (this can also be referred to as NO ROTATIONAL SYMMETRY). These shapes only look the same as their original position when they have been rotated one complete turn (360°).

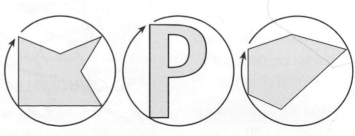

Scale Drawings & Map Scales

Drawing to Size and Scale

Before you attempt to construct any drawing to a particular size or scale you need the following:

- A PENCIL so any mistakes can be rubbed out.
- A RULER for the drawing of all straight lines.

- A PROTRACTOR for the measuring of any angles.
- A COMPASS for drawing circles, arcs and for any constructions.

Example

Here is a sketch map of an island. The map has four marker points A, B, C and D.

a) Make an accurate scale drawing of the quadrilateral ABCD using a scale of 1cm to represent 10km.

b) Calculate the length of BC.

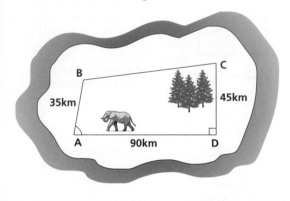

a) **Step 1: Draw the 90km line (9cm = 90km).**
Step 2: Measure and mark 80° and 90°.
Step 3: Draw the 35km line (3.5cm = 35km) and the 45km line (4.5cm = 45km).
Step 4: Complete the quadrilateral.

b) **Length of BC = 8.4cm = 8.4 x 10km = 84km**

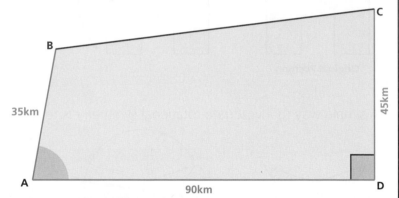

Map Scales

This is part of a map of Devon and Cornwall which is drawn to a scale of 1cm : 10km

1 The direct distance from Newquay to Plymouth as measured on the map is 7cm. Calculate the actual distance.

> Actual dist = Map dist x 10

> = 7 x 10 = 70km

2 The actual direct distance between Torquay and Exeter is 30km. Calculate the map distance.

> Map dist = Actual dist ÷ 10

> $= \frac{30}{10} = 3cm$

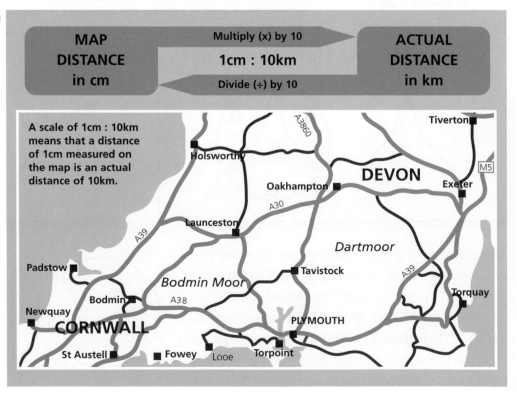

A scale of 1cm : 10km means that a distance of 1cm measured on the map is an actual distance of 10km.

Measuring Bearings

Three-Figure Bearings

A bearing is a measurement of the position of one point relative to another point. It is measured in degrees. **Bearings are always measured from the north in a clockwise direction and are given as 3 digits**. Below are two points, A and B. There are TWO possible bearings:

- The bearing of B from point A. This means that the measurement of the bearing is taken from point A.
- The bearing of A from point B. This means that the measurement of the bearing is taken from point B.

A circular protractor with a full 360° range can make three-figure bearings much easier to measure.

Sometimes the angle you measure from the N direction is less than 100° or even less than 10°. In this case one or two zeros are put in front of the angle in order to make them THREE-FIGURE BEARINGS.

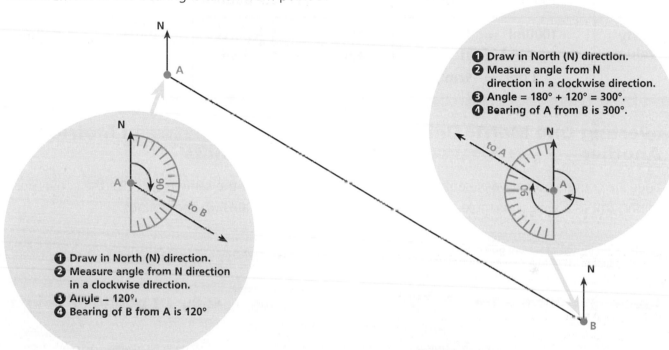

❶ Draw in North (N) direction.
❷ Measure angle from N direction in a clockwise direction.
❸ Angle = 120°.
❹ Bearing of B from A is 120°

❶ Draw in North (N) direction.
❷ Measure angle from N direction in a clockwise direction.
❸ Angle = 180° + 120° = 300°.
❹ Bearing of A from B is 300°.

Examples...

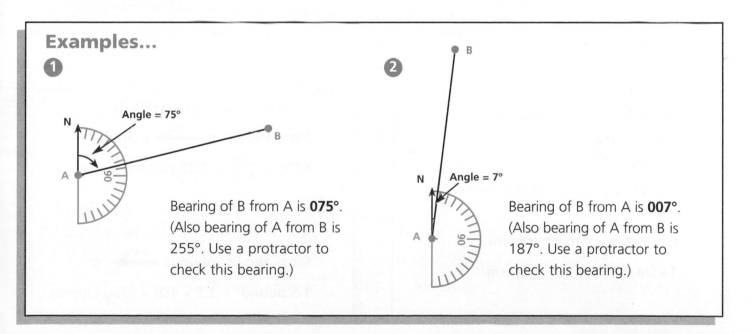

1

Angle = 75°

Bearing of B from A is **075°**.
(Also bearing of A from B is 255°. Use a protractor to check this bearing.)

2

Angle = 7°

Bearing of B from A is **007°**.
(Also bearing of A from B is 187°. Use a protractor to check this bearing.)

Converting Measurements

Metric and Imperial Units

	Metric units		Approximate comparison between Metric & Imperial		Imperial units	
Length	10mm	= 1cm	2.5cm	≈ 1 inch	12 inches	= 1 foot
	100cm	= 1m	1m	≈ 39 inches	3 feet	= 1 yard
	1000m	= 1km	1600m	≈ 1 mile	1760 yards	= 1 mile
			8km	≈ 5 miles		
Mass	1000mg	= 1g	30g	≈ 1 ounce	16 ounces	= 1 pound
	1000g	= 1kg	450g	≈ 1 pound	14 pounds	= 1 stone
	1000kg	= 1 tonne	1kg	≈ 2.2 pounds		
Capacity or Volume	1000ml	= 1l	1l	≈ $1\frac{3}{4}$ pints	8 pints	= 1 gallon
	1000cm³	= 1l	4.5l	≈ 1 gallon		
	(1ml	= 1cm³)				

Converting One Metric Unit to Another

Consider the conversion between centimetres (cm) and metres (m) as a typical example:

> Divide (÷) by 100 as we are going from a bigger number (100) to a smaller number (1)
>
> **100cm** → **100cm = 1m** → **1m**
>
> Multiply (x) by 100 as we are going from a smaller number (1) to a bigger number (100)

Examples...

① Convert 300cm to metres.

From above: cm $\xrightarrow{\div 100}$ m

$300\text{cm} = \dfrac{300}{100} = 3\text{m}$

② Penny is 1.65m tall. What is her height in centimetres?

From above: m $\xrightarrow{\text{x }100}$ cm

$1.65\text{m} = 1.65 \times 100 = 165\text{cm}$

Converting Between Metric and Imperial Units

These follow the same rules. Take the conversion between grams (g) and pounds.

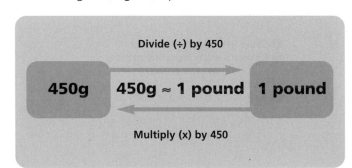

> Divide (÷) by 450
>
> **450g** → **450g ≈ 1 pound** → **1 pound**
>
> Multiply (x) by 450

Examples...

① A tin of baked beans has a mass of 600g. What is its mass in pounds?

From above: g $\xrightarrow{\div 450}$ pounds

$600\text{g} = \dfrac{600}{450} = 1.33$ **pounds** (approx.)

② A recipe for a cake needs 1.5 pounds of flour. What mass of flour is needed in grams?

From above: pounds $\xrightarrow{\text{x }450}$ g

1.5 pounds = 1.5 x 450 = 675g (approx.)

Compound Measures

Speed

This is a measure of 'how fast' an object is moving. To calculate the speed of a moving object we need two measurements:

- the **DISTANCE** it moves and
- the **TIME** it takes to move that distance.

Speed can be calculated using the formula:

$$\text{Speed (S)} = \frac{\text{Distance (D)}}{\text{Time (T)}}$$

Speed is measured - in metres per second, m/s
- or kilometres per hour, km/h
- or miles per hour, mph

A FORMULA TRIANGLE makes it a lot easier for us when we want to calculate distance or time.

To get the formula for DISTANCE, cover 'D' up

DISTANCE = SPEED x TIME

To get the formula for TIME, cover 'T' up

$$\textbf{TIME} = \frac{\textbf{DISTANCE}}{\textbf{SPEED}}$$

Examples...

1 Calculate the speed of a car which travels a distance of 90m in 10s.

$$\textbf{Speed} = \frac{\textbf{90m}}{\textbf{10}} = \textbf{9m/s} \text{ (remember the units)}$$

2 A train completes a journey of 150km at an average speed of 90km/h. How long did the journey take?

$$\textbf{Time} = \frac{\textbf{Distance}}{\textbf{Speed}} = \frac{\textbf{150km}}{\textbf{90km/h}}$$

$$= 1.\dot{6} = 1\frac{2}{3} = \textbf{1 hour 40 minutes}$$
(remember the units)

Density

This is a measure of 'how heavy' an object is 'per unit volume'. To calculate the density of an object we need two measurements:

- its **MASS** and
- its **VOLUME**.

Density can be calculated using the formula:

$$\text{Density (D)} = \frac{\text{Mass (M)}}{\text{Volume (V)}}$$

Density is measured
- in grams per centimetre cubed, g/cm³
- or kilograms per metre cubed, kg/m³

A FORMULA TRIANGLE makes it a lot easier for us when we want to calculate mass or volume.

To get the formula for MASS, cover 'M' up

MASS = DENSITY x VOLUME

To get the formula for VOLUME, cover 'V' up

$$\textbf{VOLUME} = \frac{\textbf{MASS}}{\textbf{DENSITY}}$$

Example...

Calculate the density of an object which has a mass of 75g and a volume of 100cm³.

$$\textbf{Density} = \frac{\textbf{75g}}{\textbf{100cm}^3} = \textbf{0.75g/cm}^3 \text{ (remember the units)}$$

Constructions 1

Construction of Triangles

In your exams, you must show all construction lines. If you are given…

… 3 sides: Construct the triangle ABC where AB = 3cm, BC = 2.5cm and AC = 2cm

Draw AB of length 3cm

Draw arc from A of radius 2cm

Draw arc from B of radius 2.5cm

Complete the triangle

… 2 sides & the included angle: Construct the triangle ABC where AB = 3cm, AC = 2cm and BÂC = 70°

Draw AB of length 3cm

At A, measure and mark angle of 70°

Draw AC of length 2cm

Complete the triangle

… 2 sides & the non-included angle: Construct the triangle ABC where AB = 2cm, AC = 2cm and AB̂C = 50°

Draw AB of length 2cm

At B, measure, mark and draw angle of 50°

Draw arc from A of radius 2cm

Complete the triangle

… 1 side & 2 angles: Construct the triangle ABC where AB = 3cm, BÂC = 50° and AB̂C = 30°

Draw AB of length 3cm

At A, measure, mark and draw angle of 50°

At B, measure and mark angle of 30°

Complete the triangle

Construction of an Angle of 60° and 90°

Angle of 60°

Draw arc from A to cross AB at P

Draw another arc from A of same radius

Draw arc from P (again same radius)

Complete the angle

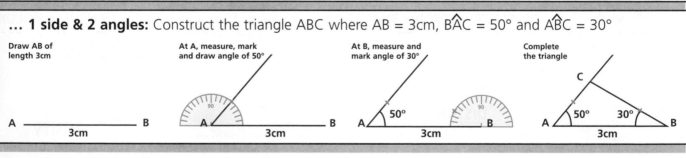

Angle of 90°

Extend A with a dotted line to C. Draw two arcs from A (same radius) to cross CB at P and Q

Draw arc from P of longer radius than in previous diagram

Draw arc from Q (same radius as previous diagram)

Complete the angle

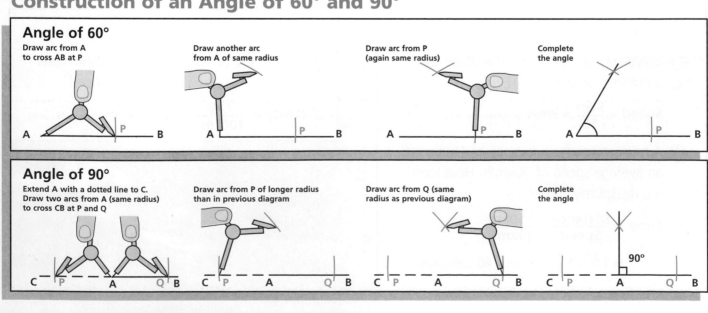

The Midpoint and Perpendicular Bisector of a Line Segment

Draw arcs of equal radius from points A and B to intersect at C.

Draw arcs of the same radius on the other side of the line to intersect at D.

Join C to D to form the perpendicular bisector, (or to locate the midpoint).

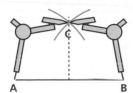

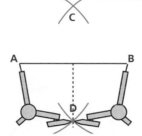

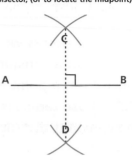

The Perpendicular from a Point to a Line

Draw TWO arcs from O to cross AB at P and Q.

Draw arc from R.

Draw arc from Q (using same radius as previous diagram).

Complete the perpendicular.

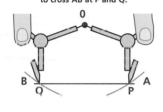

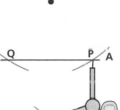

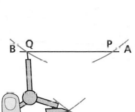

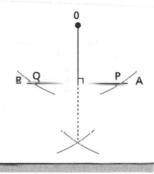

The Perpendicular from a Point on a Line

Draw arcs of equal radius from point 0, to cross AB at P and Q.

Draw arcs of the same radius (but greater than step 1) from P and Q to intersect at R.

Join 0 to R to form the perpendicular from point 0.

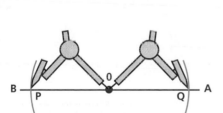

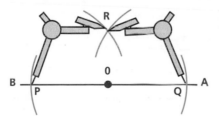

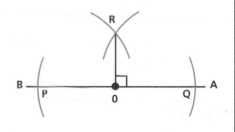

The Bisector of an Angle

Draw arcs of equal radius from point A to cut lines at B and C.

Draw arcs of equal radius from points B and C to intersect at point D.

Join A to D to form the bisector of the angle.

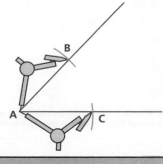

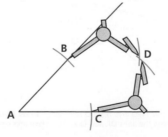

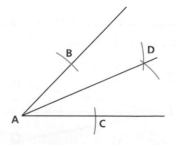

Loci

A locus is a line, all the points of which follow a particular rule. You need to know the following loci. The rule of each locus is in the box alongside the diagrams.

The locus of points which are always at a constant distance from a point is a CIRCLE whose radius is equal to the constant distance.	
The locus of points which are always at a constant distance from a line is a PAIR OF PARALLEL LINES, one above and one below the line with a PAIR OF SEMI-CIRCLES, one at each end of the line.	
The locus of points which are always equidistant from two points is a LINE WHICH BISECTS THE TWO POINTS AT RIGHT ANGLES (i.e. a perpendicular bisector). P ●--------□--------● Q	 Draw TWO arcs from P Draw TWO arcs from Q (same radius as from P) Draw the line from one intersection of arcs to the other.
The locus of points which are always equidistant from two diverging lines is a LINE WHICH BISECTS THE ANGLE BETWEEN THE TWO LINES (i.e. an angle bisector).	 Draw TWO arcs from O to cross the lines OP and OQ Draw TWO arcs from where the first two arcs cross the lines to form point R. Draw the line from R to O to form the angle bisector.

Properties of Circles 1

The following nine diagrams relate to terms used to describe various properties of circles. You need to be completely familiar with all of these.

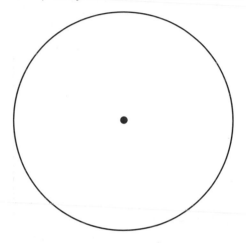

The **CENTRE** of the circle is the only point which is the same distance from every point on the circumference.

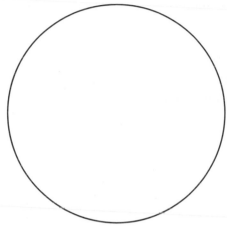

The **CIRCUMFERENCE** is the line which defines the edge of the circle. It is always the same distance from the centre of the circle.

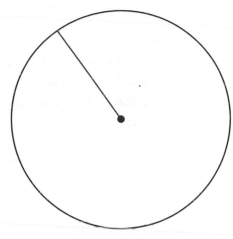

The **RADIUS** is formed by any straight line drawn from the centre of the circle to the circumference. It is always half the diameter of the circle.

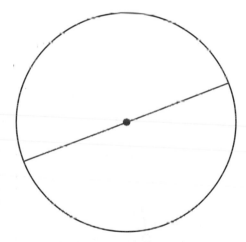

The **DIAMETER** is a straight line which passes through the centre of the circle to join opposite points on the circumference.

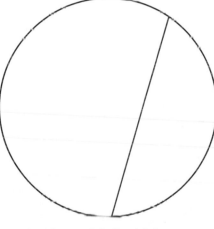

A **CHORD** is a straight line joining two points on the circumference which **DOES NOT** pass through the centre of the circle.

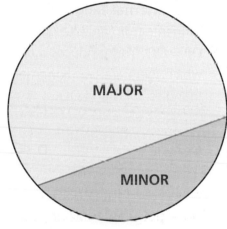

SEGMENTS are formed by a chord. The larger segment is called the major segment and the smaller one is called the minor segment.

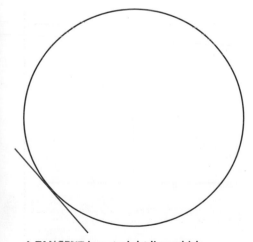

A **TANGENT** is a straight line which touches the circumference of a circle.

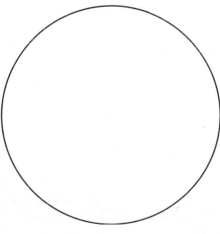

An **ARC** is simply part of the circumference of a circle.

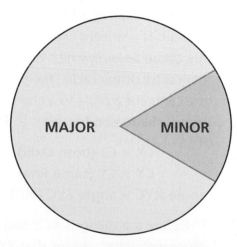

A **SECTOR** is the area enclosed by two radii and an arc. It can be major or minor.

Properties of Circles 2

There are several useful properties relating to chords and tangents of circles. You need to understand the first two of these on this page, but you must understand AND be able to explain the third.

1 Tangent and Radius

A tangent is a straight line which touches the circumference of a circle. A tangent at any point on a circle is perpendicular to the radius at that point.

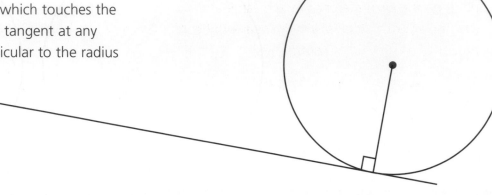

2 Two External Tangents

Tangents from an external point are equal in length, OP = OQ. This produces a symmetrical situation in which there are two congruent, right-angled triangles, OPC and OQC.

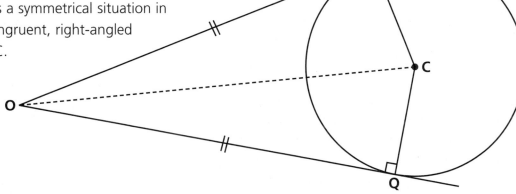

3 Perpendicular to a Chord

A chord is a straight line joining two points on the circumference, which does not pass through the centre of the circle. The perpendicular from the centre of a circle to a chord bisects the chord. The reason for this is as follows:

CX = CZ (both radii)
CY = CY (same line)
angle XYC = angle ZYC (90°)

Therefore the triangles XCY and ZCY are congruent, which means that XY = ZY

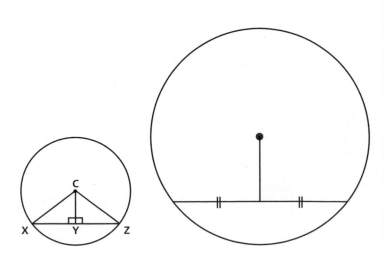

Properties of Circles 3

The following two pages have five more facts relating to circles and angles. You need to know them and their proofs.

1 Angles subtended by an arc.

The angle at the centre of a circle, subtended by an arc, is TWICE the angle subtended at any point on the circumference by the same arc. (NB The word subtended simply means to be produced by drawing straight lines.)

> **Angle AOB = 2 x Angle ACB**

> **Reflex Angle AOB = 2 x Angle ADB**

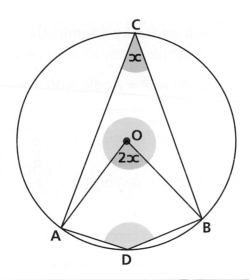

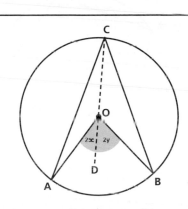

Let $A\hat{O}D = 2x$ and $B\hat{O}D = 2y$

$A\hat{O}C = 180° - 2x$ (Angles on a straight line) and

$A\hat{C}O = \frac{180° - A\hat{O}C}{2} = \frac{180° - (180° - 2x)}{2} = \frac{2x}{2} = x$ (isosceles triangle)

Similarly, $B\hat{O}C = 180° - 2y$ (Angles on a straight line) and

$B\hat{C}O = \frac{180° - B\hat{O}C}{2} = \frac{180° - (180° - 2y)}{2} = \frac{2y}{2} = y$ (isosceles triangle)

$\therefore A\hat{O}B = 2x + 2y$ and $A\hat{C}B = A\hat{C}O + B\hat{C}O = x + y$
which means that $A\hat{O}B = 2 \times A\hat{C}B$

2 Angles subtended in the same segment.

Angles subtended in the same segment are all equal. Here the angles subtended by the chord AB in the major segment are equal. The angles subtended by the chord AB in the minor segment are also equal.

> **Angle ACB = Angle ADB**

> **Angle AEB = Angle AFB**

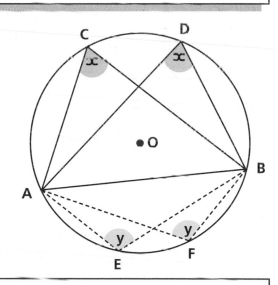

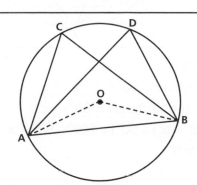

$A\hat{O}B = 2 \times A\hat{C}B$ (Angles subtended by an arc - see **1**)
$A\hat{O}B = 2 \times A\hat{D}B$ (Angles subtended by an arc - see **1**)
$\therefore A\hat{C}B = A\hat{D}B$

Properties of Circles 4

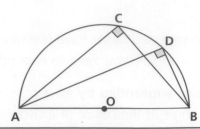

3 **Angles subtended by a semicircle**

The angles subtended by a semicircle are always 90°.

> **Angle ACB = Angle ADB = 90°**

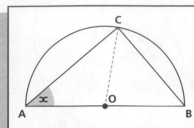

Let $\hat{CAO} = x$ and so $\hat{ACO} = x$ (Isosceles triangle)

$\therefore \hat{AOC} = 180° - (\hat{CAO} + \hat{ACO}) = 180° - 2x$

$\hat{BOC} = 180° - \hat{AOC} = 180° - (180° - 2x) = 2x$ (Angles on a straight line)

$\therefore \hat{BCO} = \frac{180° - \hat{BOC}}{2} = \frac{180° - 2x}{2} = 90° - x$ (Isosceles triangle)

$\hat{ACB} = \hat{ACO} + \hat{BCO} = x + (90° - x) = 90°$

4 **Opposite angles of a cyclic quadrilateral**

Opposite angles of a cyclic quadrilateral add up to 180°. A cyclic quadrilateral is a quadrilateral whose vertices (corners) all lie on the circumference of the same circle. In the example shown alongside…

> **Angle BAD + Angle BCD = 180°**

> **Angle ABC + Angle ADC = 180°**

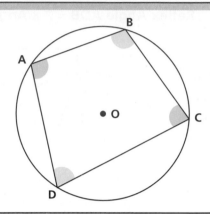

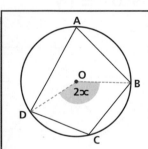

Let $\hat{BOD} = 2x$ $\therefore \hat{BAD} = \frac{\hat{BOD}}{2} = \frac{2x}{2} = x$ (Angles subtended by an arc)

Reflex $\hat{BOD}$ = 360° − 2x (Angles at a point)

$\therefore \hat{BCD} = \frac{\text{Reflex } \hat{BOD}}{2} = \frac{360° - 2x}{2} = 180° - x$ (Angles subtended by an arc)

$\hat{BAD} + \hat{BCD} = x + (180° - x) = 180°$

Consequently $\hat{ABC} + \hat{ADC} = 180°$ (Angles of a quadrilateral = **360°**)

5 **Alternate segment theorem**

The angle subtended between a tangent to a circle and its chord is equal to the angle subtended in the alternate segment.

> **$\hat{BAE} = \hat{ACB} = \hat{ADB}$**

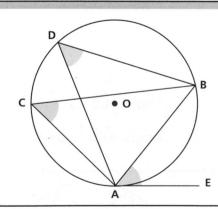

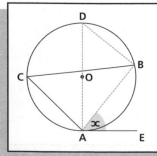

Let $\hat{BAE} = x$ and triangle **ABD** have **AD** as diameter.

$\hat{ABD} = 90°$ (Angle subtended by a semicircle = **90°**)

$\hat{BAD} = 90° - x$ (Angle between tangent and radius = **90°**)

$\therefore \hat{ADB} = 180° - (\hat{BAD} + \hat{ABD}) = 180° - [(90° - x) + 90°] = 180° - 90° + x - 90° = x$

$\hat{ACB} = \hat{ADB}$ (Angles subtended by an arc)

and consequently $\hat{ACB} = \hat{BAE}$

Probability 1

The Nature of Probability

Probability, quite simply is a measure of the likelihood of a particular event occurring. Probability is used throughout the world in order to give a sound mathematical basis for predicting future events. These events can be relatively simple such as tossing a coin or spinning a roulette wheel, however they can also be vastly complicated such as in the assessment of a large insurance risk.

Probability Scale and the Language of Probability

For an event to happen we might say that its probability is 'a certainty', 'more than likely', 'evens' or '50/50', 'not very likely' or 'no chance'. The probability of an event occurring (or not occurring) can be stated as either a fraction, percentage or decimal. These can be shown on a probability scale. Here is a very simple one:

These are the only probabilities you can have, i.e. you cannot have a probability greater than 1.
So, if the probability of something happening is P, then the probability of it not happening is 1 − P

0	$\frac{1}{10}$	$\frac{2}{10}$	$\frac{3}{10}$	$\frac{4}{10}$	$\frac{5}{10}$	$\frac{6}{10}$	$\frac{7}{10}$	$\frac{8}{10}$	$\frac{9}{10}$	1
0	10%	20%	30%	40%	50%	60%	70%	80%	90%	100%
0	0.1	0.2	0.3	0.4	0.5	0.6	0.7	0.8	0.9	1

| No chance i.e. impossible | | Not very likely | | Evens 50/50 | | More than likely | | A certainty i.e. bound to happen |

Examples...

1 The probability of you passing a maths test is $\frac{7}{10}$. What is the probability that you will fail?

> Since you can only pass or fail, the total probability = 1. Therefore...

P(passing) + P(failing) = 1

P(failing) = 1 − P(passing)

$$= 1 - \frac{7}{10}$$

$$= \frac{3}{10}$$

> ... which means that you are not very likely to fail!

2 A drawer contains white, black and red socks only. If the probability of picking a white sock at random is 0.2 and a black sock is 0.3, what is the probability of picking a red sock?

> Since the drawer only contains white, black and red socks, the total probability = 1. Therefore...

P(w) + P(b) + P(r) = 1

P(r) = 1 − (P(w) + P(b))

= 1 − (0.2 + 0.3)

= 1 − 0.5

= 0.5

> ... which means that you have an even chance of picking a red sock at random.

Probability 2

When either one outcome or the other outcome can occur, the two outcomes are said to be MUTUALLY EXCLUSIVE. This can also be the case when there are three or more possible outcomes.

Outcomes with Equal Chances

The important rule to remember is this: 'If there are 'n' mutually exclusive outcomes, all of which are equally likely, then the probability of one outcome happening is $\frac{1}{n}$'.

Examples...

1 **Tossing a coin:** Here there are two mutually exclusive outcomes, Heads or Tails - if you throw a Head you can't throw a Tail.

P(Heads) = $\frac{1}{2}$
P(Tails) = $\frac{1}{2}$

Note that the probability of each outcome added together is 1, i.e. $\frac{1}{2} + \frac{1}{2} = 1$

2 **Spinning a Spinner:** Here there are three mutually exclusive outcomes since if you spin a red, you can't spin any other colour!

P(Red) = $\frac{1}{3}$
P(Blue) = $\frac{1}{3}$
P(Green) = $\frac{1}{3}$

Note that the probability of each outcome added together is 1, i.e. $\frac{1}{3} + \frac{1}{3} + \frac{1}{3} = 1$

3 **Throwing a die:** Here are six mutually exclusive outcomes, since throwing any one number prevents you from throwing the rest.

P(one) = $\frac{1}{6}$
P(two) = $\frac{1}{6}$
P(three) = $\frac{1}{6}$
P(four) = $\frac{1}{6}$
P(five) = $\frac{1}{6}$
P(six) = $\frac{1}{6}$

Note that the probability of each outcome added together is 1, i.e. $\frac{1}{6} + \frac{1}{6} + \frac{1}{6} + \frac{1}{6} + \frac{1}{6} + \frac{1}{6} = 1$

Outcomes with Unequal Chances

The important rule to remember is this: 'If there are 'n' mutually exclusive outcomes, but only 'a' desired outcomes, then the probability of a desired outcome is $\frac{a}{n}$'.

Examples...

1 **Picking a Ball (from inside a bag):** With this set of balls there are four mutually exclusive outcomes - black, green, yellow and orange. However, the chances of each colour being picked are not equal, since there are 5 blacks, 4 greens, 2 yellows and 1 orange.

P(black) = $\frac{5}{12}$
P(green) = $\frac{4}{12}$ or $\frac{1}{3}$
P(yellow) = $\frac{2}{12}$ or $\frac{1}{6}$
P(orange) = $\frac{1}{12}$

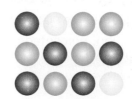

Note that the probability of each outcome added together is 1, i.e. $\frac{5}{12} + \frac{1}{3} + \frac{1}{6} + \frac{1}{12} = 1$

2 **Spinning a Spinner:** With this spinner there are three mutually exclusive outcomes - red, blue and green. Since the spinner has 3 reds, 2 blues and 1 green, there is an unequal chance of getting each colour.

P(red) = $\frac{3}{6}$ or $\frac{1}{2}$
P(blue) = $\frac{2}{6}$ or $\frac{1}{3}$
P(Green) = $\frac{1}{6}$

Note that the probability of each outcome added together is 1, i.e. $\frac{1}{2} + \frac{1}{3} + \frac{1}{6} = 1$

Probability 3

Addition of Probabilities for Mutually Exclusive Outcomes

The important rule to remember is this: 'If **A** and **B** are mutually exclusive outcomes (or events), then the probability of **A or B** occurring is **P(A) + P(B)**, for example…

With this spinner there are three mutually exclusive outcomes - red, blue and green. Since the spinner has 3 reds, 2 blues and 1 green, there is an unequal chance of getting each colour.

$P(red) = \frac{3}{6}$ or $\frac{1}{2}$
$P(blue) = \frac{2}{6}$ or $\frac{1}{3}$
$P(Green) = \frac{1}{6}$

And so, the probability of getting a red **or** a blue is
$P(red) + P(blue) = \frac{1}{2} + \frac{1}{3} = \frac{5}{6}$

Similarly the probability of getting a red **or** a green is
$P(red) + P(green) = \frac{1}{2} + \frac{1}{6} = \frac{2}{3}$

And the probability of getting a blue **or** a green is
$P(blue) + P(green) = \frac{1}{3} + \frac{1}{6} = \frac{1}{2}$

Multiplication of Probabilities for Independent Outcomes

The important rule to remember is this: 'If **A** and **B** are independent outcomes (i.e. one event does not depend on the other), then the probability of **A and B** occurring is **P(A) x P(B)**, for example…

If you toss a coin and spin this spinner then tossing a coin gives two mutually exclusive outcomes and the spinner gives three mutually exclusive outcomes. However they are independent events.

$P(Heads) = \frac{1}{2}$ $P(red) = \frac{1}{3}$
$P(Tails) = \frac{1}{2}$ $P(blue) = \frac{1}{3}$
$P(green) = \frac{1}{3}$

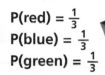

And so, the probability of getting a Head (with the coin) **and** a red (with the spinner) is…
$P(Head) \times P(red) = \frac{1}{2} \times \frac{1}{3} = \frac{1}{6}$

Similarly there are five other possible outcomes, which are fairly straightforward. You should find that the probability of each outcome is $\frac{1}{6}$.

Examples…

1 What is the probability of throwing a 1 or a 2 with a die and picking a red or a green ball from inside a bag which contains 4 black, 5 red and 3 green balls? Throwing the die…

$P(1)$ or $P(2) = \frac{1}{6} + \frac{1}{6} = \frac{2}{6} = \frac{1}{3}$
These are mutually exclusive outcomes and so we **add** the probabilities.

Picking a red or a green ball…
$P(red)$ or $P(green) = \frac{5}{12} + \frac{3}{12} = \frac{8}{12} = \frac{2}{3}$
These are mutually exclusive outcomes and so we **add** the probabilities.

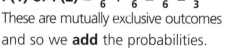

Therefore,
P(1 or 2) and **P(red or green)** $= \frac{1}{3} \times \frac{2}{3} = \frac{2}{9}$

These are independent outcomes and so we **multiply** the probabilities.

2 A bag contains 4 black, 5 red and 3 green balls. What is the probability of picking one ball and then, without replacing the ball, picking another of the same colour from the bag?

 The important thing to remember is that if you pick a black ball then the number of black balls in the bag is one less etc.

Firstly we need to work out the probability of independently picking 2 blacks, 2 reds and 2 greens.

$P(B) + P(B) = \frac{4}{12} \times \frac{3}{11} = \frac{12}{132} = \frac{1}{11}$
<small>1st pick   2nd pick</small>

$P(R) + P(R) = \frac{5}{12} \times \frac{4}{11} = \frac{20}{132} = \frac{5}{33}$
<small>1st pick   2nd pick</small>

$P(G) + P(G) = \frac{3}{12} \times \frac{2}{11} = \frac{6}{132} = \frac{1}{22}$
<small>1st pick   2nd pick</small>

Therefore, the probability of picking
P(BB) or **P(RR)** or **P(GG)** $= \frac{1}{11} + \frac{5}{33} + \frac{1}{22} = \frac{19}{66}$
These are mutually exclusive outcomes and so we **add** the probabilities.

Listing All Outcomes

Outcomes of Single Events

1 **Tossing a coin.** Here there are only two outcomes: Heads or Tails

2 **Spinning a three sided spinner.** Here there are three outcomes: Red, Blue or Green

3 **Throwing a die.** Here there are only six outcomes: One, Two, Three, Four, Five or Six

In all cases like these all you have to do is make a simple list.

Outcomes of Two Successive Events

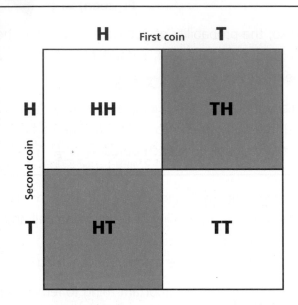

1 **Tossing two coins, one after the other.**
In cases like these it is easier to use a sample space diagram to show the possible outcomes. With two coins there are 4 outcomes. The chances of each are…

P(Head + Head) = $\frac{1}{4}$
P(Tail + Tail) = $\frac{1}{4}$
P(Head + Tail) = $\frac{1}{4}$
P(Tail + Head) = $\frac{1}{4}$

Note that once again the probability of each outcome added together is 1, i.e. $\frac{1}{4} + \frac{1}{4} + \frac{1}{4} + \frac{1}{4} = 1$

2 **Throwing a pair of dice.**
In this case, each die has six possible outcomes. The sample space diagram, which shows the scores of the two dice added together, reveals a total of 36 possible outcomes. Each outcome (e.g. throwing a 4 with the first die and a 2 with the second) has a $\frac{1}{36}$ probability.

However, while there is only a $\frac{1}{36}$ chance of scoring a total of either 2 or 12, there is a $\frac{6}{36}$ or $\frac{1}{6}$ chance of scoring a seven.

Tree Diagrams 1

When dealing with the probability of a sequence of successive events it is often better to use a tree diagram in order to simplify the task.

Example 1 - Tossing Two Coins

Again each coin can give us two possible outcomes, a HEAD (H) or a TAIL (T).

We only draw two branches as there are only two different outcomes i.e. a Head and a Tail.

$$\frac{1}{2}$$ H

$$\frac{1}{2}$$ H $$\frac{1}{2}$$ T

$$\frac{1}{2}$$ T $$\frac{1}{2}$$ H $$\frac{1}{2}$$ T

H P(HH) = P(H) x P(H) = $\frac{1}{2}$ x $\frac{1}{2}$ = $\frac{1}{4}$

T P(HT) = P(H) x P(T) = $\frac{1}{2}$ x $\frac{1}{2}$ = $\frac{1}{4}$

H P(TH) = P(T) x P(H) = $\frac{1}{2}$ x $\frac{1}{2}$ = $\frac{1}{4}$

T P(TT) = P(T) x P(T) = $\frac{1}{2}$ x $\frac{1}{2}$ = $\frac{1}{4}$

These are independent outcomes

First Coin   **Second Coin**   Total probability always adds up to **1**

Probability of getting:

① 2 Heads, P(HH) = $\frac{1}{4}$

② 2 Tails, P(TT) = $\frac{1}{4}$

③ **A Head and a Tail, P(HT) or P(TH)**

= $\frac{1}{4}$ + $\frac{1}{4}$ = $\frac{1}{2}$

(these are mutually exclusive outcomes)

Example 2

From Jim's bag, the probability of selecting a red ball is 0.5, a blue ball is 0.3 and a green ball is 0.2. From Sandra's bag, the probability of selecting a red ball is 0.3, a blue ball is 0.3 and a green ball is 0.4. Jim selects a ball at random from his bag, followed by Sandra from hers. Draw a tree diagram to show all the different outcomes. Then calculate the probability of selecting...

① two blues.

② two reds or two greens.

③ different colours.

This time we draw three branches as there are three different outcomes

R	P (RR)	= 0.5 x 0.3 = 0.15
B	P (RB)	= 0.5 x 0.3 = 0.15
G	P (RG)	= 0.5 x 0.4 = 0.20
R	P (BR)	= 0.3 x 0.3 = 0.09
B	P (BB)	= 0.3 x 0.3 = 0.09
G	P (BG)	= 0.3 x 0.4 = 0.12
R	P (GR)	= 0.2 x 0.3 = 0.06
B	P (GB)	= 0.2 x 0.3 = 0.06
G	P (GG)	= 0.2 x 0.4 = 0.08

These are independent outcomes

Jim's Bag   **Sandra's Bag**   Total probability always adds up to **1**

① P(BB) = 0.09

② P(RR) or P(GG) = 0.15 + 0.08 = 0.23
(these are mutually exclusive outcomes)

③ P(different colours)
= P(RB), P(RG), P(BR), P(BG), P(GR), P(GB)
= 0.15 + 0.20 + 0.09 + 0.12 + 0.06 + 0.06
= **0.68** (these are mutually exclusive outcomes)

Tree Diagrams 2

Sometimes the probability of an event occurring depends on the outcome of a previous event. For example, a question may ask you to remove an object at random and then remove a second object at random without replacing the first. When this happens you must remember to adjust your probabilities for the second event.

Example

Jack's bag contains 10 balls of which 5 are red, 3 are blue and 2 are green. Jack picks a ball at random and, without replacing it, picks another ball at random.

Draw a tree diagram to show all the different possible outcomes. Use the diagram to calculate the probability of Jack picking balls that are…

1 the same colour.

2 different colours.

- You know that after the first pick the ball is not replaced in the bag. Therefore there are only 9 balls in Jack's bag for the second pick.

- If for example, the first pick is a red ball, then when the second pick is made there are only 4 red balls in the bag.

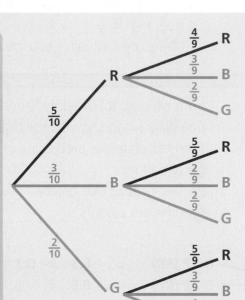

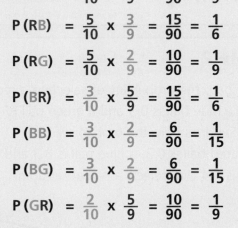

$$P(RR) = \frac{5}{10} \times \frac{4}{9} = \frac{20}{90} = \frac{2}{9}$$

$$P(RB) = \frac{5}{10} \times \frac{3}{9} = \frac{15}{90} = \frac{1}{6}$$

$$P(RG) = \frac{5}{10} \times \frac{2}{9} = \frac{10}{90} = \frac{1}{9}$$

$$P(BR) = \frac{3}{10} \times \frac{5}{9} = \frac{15}{90} = \frac{1}{6}$$

$$P(BB) = \frac{3}{10} \times \frac{2}{9} = \frac{6}{90} = \frac{1}{15}$$

$$P(BG) = \frac{3}{10} \times \frac{2}{9} = \frac{6}{90} = \frac{1}{15}$$

$$P(GR) = \frac{2}{10} \times \frac{5}{9} = \frac{10}{90} = \frac{1}{9}$$

$$P(GB) = \frac{2}{10} \times \frac{3}{9} = \frac{6}{90} = \frac{1}{15}$$

$$P(GG) = \frac{2}{10} \times \frac{1}{9} = \frac{2}{90} = \frac{1}{45}$$

These are independent outcomes

| First Pick | Second Pick | Total probability always adds up to **1** |

1 P(same colour)

= P(RR) or P(BB) or P(GG)

$= \frac{2}{9} + \frac{1}{15} + \frac{1}{45}$

$= \frac{14}{45}$

(these are mutually exclusive outcomes)

2 P(different colour)

= P(RB) or P(RG) or P(BR) or P(BG) or P(GR) or P(GB)

$= \frac{1}{6} + \frac{1}{9} + \frac{1}{6} + \frac{1}{15} + \frac{1}{9} + \frac{1}{15}$

$= \frac{31}{45}$

(these are mutually exclusive outcomes)

Theoretical Probability

Since the chance of a coin landing on Heads is $\frac{1}{2}$, then the number of Heads we should expect in 10 tosses is…

$\frac{1}{2}$ **x 10 = 5**.

This however is not always the case in reality:

Estimated Probability (Relative Frequency)

A simple experiment was carried out where a coin was tossed 10, 100 and a 1000 times. The graphs opposite show the number of Heads and Tails obtained. In our experiment we would expect to always get the same number of Heads and Tails as the probability of each event occurring is $\frac{1}{2}$ or 0.5.

The actual probability we get when we perform an experiment like this is called the RELATIVE FREQUENCY.

Relative Frequency =

$$\frac{\textbf{Number of Heads (or Tails) we get}}{\textbf{Total number of times the coin was tossed}}$$

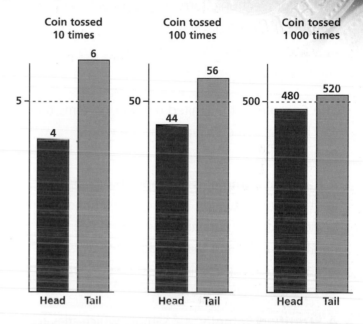

If we go back to our experiment, as we increase the number of times the coin is tossed, the relative frequency gets closer and closer to the theoretical probability (red dotted line), e.g. 0.5 for a Head, 0.5 for a Tail.

	Coin tossed 10 times	Coin tossed 100 times	Coin tossed 1 000 times
Relative Frequency	Head $= \frac{4}{10} = 0.4$	Head $= \frac{44}{100} = 0.44$	Head $= \frac{480}{1000} = 0.48$
	Tail $= \frac{6}{10} = 0.6$	Tail $= \frac{56}{100} = 0.56$	Tail $= \frac{520}{1000} = 0.52$

Importance of Sample Size

As you can see from the data above, the bigger the sample, the more reliable it is. It would be easy to believe that tossing a coin ten times could produce 2 tails and 8 heads!

This has implications in testing for **bias**. For instance it would be possible to test experimentally the frequency of red and black on a roulette wheel. If there was time to take a big enough sample then it would be possible to say with some justification that the wheel wasn't fair, e.g. if 50 000 spins produce 26 750 reds and 23 250 blacks!!

Collecting Data 1

Some Definitions

- **Primary Data:** is data which has been directly obtained first hand, either by yourself or by someone under your direction. It can be collected by questionnaire, survey, observation, experiment or data logging.

- **Secondary Data:** is data which has been obtained independently by an external agency and which may be already stored either in printed or electronic form, e.g. published statistics, data from the internet.

- **The Population:** is the total number of items you are investigating.

- **The Sample:** is the limited number of items that you have selected to represent the whole population.

- **A Random Sample:** is a sample in which every item of the population has an equal chance of being selected. In practice this means taking great care to spread the 'randomness' over as large an area as possible and to take repeated surveys and work out averages.

- **A Stratified Sample:** is a sample which tries to accurately reflect the variation of characteristics within the population, e.g. it reflects the sex distribution, the age distribution, the geographical distribution, the income distribution etc. It is a small accurate snapshot of the population as a whole. Done properly this method can be very accurate.

Reasons for Sampling

Information is a hugely powerful tool in modern society but clearly it is impossible to survey huge populations. Sampling allows us to look at a cross-section of the population, making the process much quicker and much cheaper! Samples are used widely to inform opinion polls, market research, to produce T.V. viewing figures and to provide trend analysis.

Identifying Bias in Samples

Consider the following bad examples of sampling:

- A recent survey suggests that 82% of the population prefer Rugby League to Soccer. The survey was conducted in St Helens.

- A High Street survey reveals that 76% of men aged between 18 and 42 go to the pub at least once a week. The survey was conducted at 11pm.

- A telephone survey suggests that 100% of the population has at least one telephone in their house.

In order to be unbiased, every individual in the population must have an equal chance of being included in the sample. This means taking into account:

- The time of day of the survey
- The age range
- Relative affluence
- The geographical area
- Ethnicity
- Lifestyle

However, you must also remember that THE BIGGER THE SAMPLE SIZE, THE MORE REPRESENTATIVE IT'S LIKELY TO BE (assuming of course that you have minimised all the other potential areas of bias).

Collecting Data by Observation

This can be laborious and time consuming but for some things it is the best way. For instance, a traffic survey might be done in this way to reveal the volume of traffic using a bridge. You must remember to ask yourself whether the survey is being conducted at and for an appropriate time.

Collecting Data by Experiment

People involved in the Sciences use experiments to gather data to support their hypotheses. The key things to remember are that the experiment must be repeated an appropriate number of times, and also that the experiment must be capable of being repeated by someone else.

Collecting Data by Questionnaire

Questionnaires are skillfully designed forms which are used to conduct surveys of a sample of the population.

Designing a Questionnaire

Good questionnaires have the following things in common…

- They are not too long. Never more than 10 questions, but less if possible.
- They contain questions which are easily understood and do not cause confusion.
- They ask for simple, short answers, e.g. Yes/No, Like/Don't Like or Male/Female
- They avoid vague words like Tall, Old, Fast, Good etc.
- The questions do not show any bias, e.g. 'Do you prefer watching rugby or hockey?' rather than 'Do you agree that rugby is a more watchable game than hockey?'
- They only contain relevant questions.

Example

Yasmin decides to test the hypothesis that 'Parents would prefer the school holidays to be shorter' by using the following questionnaire.

This tests whether the questionnaire is relevant to this person

The answer may be affected by the size of the family!

The answer may turn out to be dependent upon the age of the person's children!

This tells us whether or not the person will see a lot of his/her children over the holidays!

This avoids asking a 'loaded question' i.e. it avoids bias

QUESTIONNAIRE

1. Do you have children of school age?
   Yes ❑ No ❑

2. How many children do you have?
   1 ❑ 2 ❑ 3 ❑ 4+ ❑

3. To which age group do they belong?
   11-13 ❑ 14-16 ❑ 17-18 ❑

4. Are you in full time employment?
   Yes ❑ No ❑

5. Do you think the school holidays are…
   Too Short? ❑
   Too Long? ❑
   Just Right? ❑

From the answers, Yasmin could…

1. Reject any responses from people who aren't parents.
2. Analyse the data to see if family size, age range and employment status affect the answers.
3. Come up with a pretty good answer to her original hypothesis.

Sorting Data 1

Data comes in many different forms. To make sense of the data it is often sorted and collated. There are two different types of data that you can record for sorting and collating.

Discrete and Continuous Data

Discrete data is data that can only have certain values. For example, the number of goals a football team can score in a match is 0, 1, 2, 3, 4, etc. They cannot have a score in between, like 0.5, 1.6, 2.2, etc.

Continuous data is data that can have any value. It tends to be obtained by reading measuring instruments. The accuracy of the data is dependent on the precision of the equipment.

Tally Charts and Frequency Tables

Very often the best way to sort and collate discrete or continuous data is to draw a tally chart and a frequency table.

Example

Here are the results of the games involving Germany, the host nation, in Round 1 through to the semi-finals of the 2006 World Cup. Sort the number of goals scored per game by each team by drawing a tally chart and a frequency table.

Since the range of data here is narrow (e.g. from 0 to 4 goals scored) each value can be included individually in the tally chart and frequency table. As you complete the tally column always tick off each number as you go along. This makes sure that you don't include the same number twice or miss any out.

The numbers in the frequency column are simply the number of tallies. Remember to add them up, as this total is equal to the total number of pieces of data (not, in this example, the total number of goals scored!).

The data in this example is discrete, however, the same process would apply for continuous data.

Germany 4̸ Costa Rica 2̸	Germany 2̸ Sweden 0̸
Germany 1̸ Poland 0̸	Germany 4̸ Argentina 2̸
Ecuador 0̸ Germany 3̸	Germany 0̸ Italy 2̸

Number of goals scored	Tally	Frequency of that no. of goals being scored by a team
0	IIII	4
1	I	1
2	IIII	4
3	I	1
4	II	2
		TOTAL = 12

Sorting Data 2

Using Class Intervals

Sometimes, unlike the example on the previous page, the data is so widespread that it is impractical to include each value in the tally chart and frequency table individually. When this happens, the data is sorted into groups called class intervals, where each class interval represents a range of values.

Example

The newspaper cutting shows the recorded temperatures in °C for various places both at home and abroad. Sort the recorded temperatures by drawing a tally chart and a frequency table.

Temperatures home and abroad

Amsterdam	19	Cairo	34	Majorca	27	New York	32
Athens	33	Cardiff	16	Manchester	12	Newcastle	13
Barbados	29	Dublin	15	Miami	26	Paris	20
Barcelona	26	Jersey	20	Milan	28	Peking	33
Berlin	23	London	21	Montreal	22	Prague	26
Bermuda	28	Madrid	33	Moscow	15	Rhodes	28

As you can see, the data here is widespread, e.g. from 12°C to 34°C. Out of practicality, the temperature values are arranged in groups of five in our tally chart and frequency table.

The class interval $10 \leqslant T < 15$ would include any temperature reading equal to or greater than 10°C and less than 15°C (a temperature reading of 15°C is included in the next class interval) and so on. Remember to total the numbers in the frequency column to make sure they add up to the total number of locations (pieces of data).

The data in this example is continuous, however, the same process would apply for discrete data.

Recorded temperatures, T(°C)	Tally	Frequency of temperatures falling within that range
$10 \leqslant T < 15$	II	2
$15 \leqslant T < 20$	IIII	4
$20 \leqslant T < 25$	IIII	5
$25 \leqslant T < 30$	IIII III	8
$30 \leqslant T < 35$	IIII	5
		TOTAL = 24

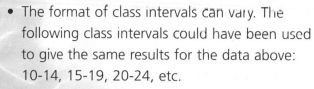

Two Important Points

- The format of class intervals can vary. The following class intervals could have been used to give the same results for the data above: 10-14, 15-19, 20-24, etc.
- The range of each class interval depends on the total range of the data. If the total range of the data is large and the range of each class interval is small, then your frequency table would have a lot of rows. Aim to have no more than 10 lines in your table!

Sorting Data 3

Stem and Leaf Diagrams

A stem and leaf diagram sorts data into groups. An advantage over frequency tables is that they enable you to get more of a feel for the 'shape' of distribution. For example, the following data shows the length of time (in minutes) it took 30 pupils to complete a test, arranged in ascending order:

8, 8, 9, 15, 15, 16, 16, 17, 18, 18, 20, 21, 21, 22, 23, 26, 27, 27, 28, 28, 29, 33, 34, 34, 35, 39, 39, 42, 48, 49.

This data can be represented using a stem and leaf diagram by taking the tens to form the 'stem' of the diagram and the units to form the 'leaves'.

The end product is similar to a frequency table. However, besides allowing you to visualise the 'shape' of the data, it can be used to identify the modal class i.e. the 20-29 group and the median, i.e. 24.5 (between the 15th and 16th piece of data as there are 30 values).

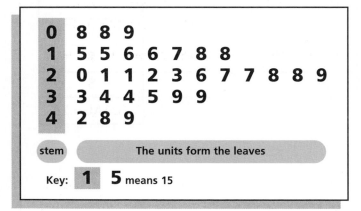

0	8 8 9
1	5 5 6 6 7 8 8
2	0 1 1 2 3 6 7 7 8 8 9
3	3 4 4 5 9 9
4	2 8 9

stem   The units form the leaves

Key: **1**   **5** means 15

Two-Way Tables

These simply show two sets of information, one vertically and the other horizontally. Information organised in this way can actually result in you gaining more information than you started with.

For instance: 'In a survey, 200 Year 7 and 8 pupils were asked if they preferred Maths or Science. 73 out of 110 year 7 pupils preferred Maths, and in total 82 pupils preferred Science.' This could lead to the table below:

	YEAR 7	YEAR 8	TOTAL
SCIENCE			82
MATHS	73		
TOTAL	110		200

… which in turn can be used to work out the missing data:

	YEAR 7	YEAR 8	TOTAL
SCIENCE	37	45	82
MATHS	73	45	118
TOTAL	110	90	200

Notice that in the original table there was no data for Year 8 … now it's all there!

Various Types of Table

Tables can be arranged in many different ways to suit the purpose for which they are intended. Remember, the idea is to make the information as accessible as possible, so you've got to give a bit of thought as to how you want to present it.

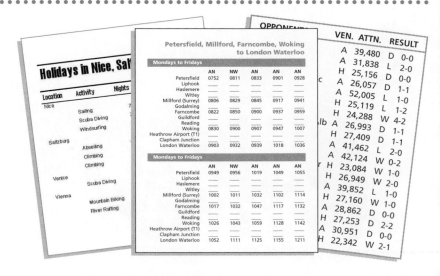

Displaying Data 1

The best way of displaying data that has been sorted into a frequency table is to draw a graph.

Here is the frequency table for the number of goals scored in games involving Germany in Round 1 through to the semi-finals of the 2006 World Cup (we only include the tally column when we are sorting the data).

No. of goals scored	Frequency
0	4
1	1
2	4
3	1
4	2

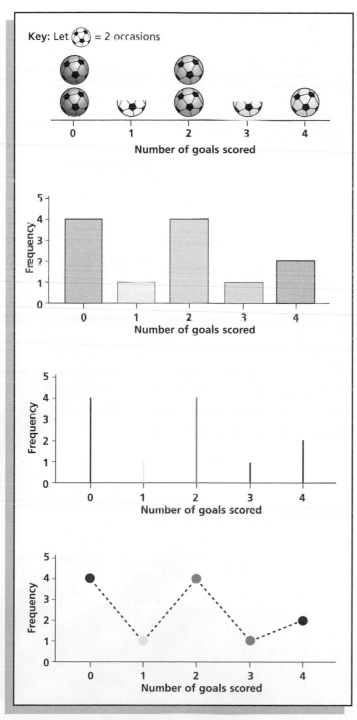

The data above can be displayed in various ways:

A Pictogram

Simple diagrams are used to display data. Since our data is about football we have used a ball. Don't forget to use the key when reading the pictogram.

A Bar Graph

Bars or columns are used to display data. Make sure that the height of each bar is equal to the correct frequency.

A Vertical Line Graph

This graph is very similar to the bar graph above, except that lines are drawn instead of bars. Make sure that the height of each line is equal to the correct frequency.

A Jagged Line Graph

This is not always the most suitable way to display discrete data, as the lines joining the points have no meaning!

Displaying Data 2

Data that has been grouped together and sorted, using class intervals, in a frequency table can also be displayed by drawing a graph.

Here is the frequency table for the recorded temperatures for various locations at home and abroad.

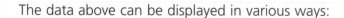

Recorded temperatures T(°C)	Frequency
10 ≤ T < 15	2
15 ≤ T < 20	4
20 ≤ T < 25	5
25 ≤ T < 30	8
30 ≤ T < 35	5

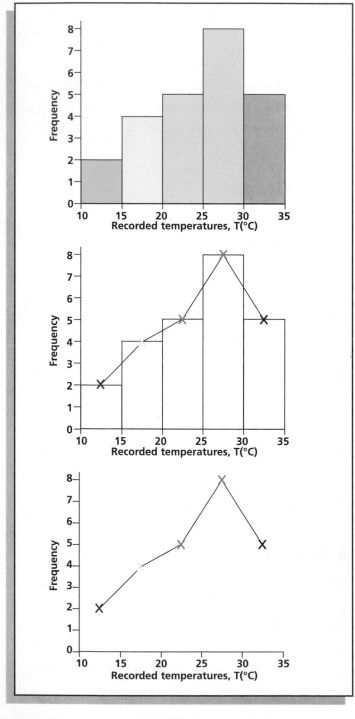

The data above can be displayed in various ways:

A Frequency Diagram

This is very much like a bar graph, but since we have grouped data then the bars do not have a gap between them, i.e. they must be continuous, one after another.

A Frequency Polygon

All you need to do is mark the middle of the top of each bar in the frequency diagram with a cross and then join up these crosses with straight lines.

You may be asked to draw a frequency polygon directly from the frequency table. You must remember to plot the crosses at the correct frequency exactly over the middle of the class intervals, e.g. for **10 ≤ T < 15**, plot the cross above 12.5

Displaying Data 3

Time Series

One of the most frequently used formats for displaying continuous data involves plotting a line graph of a particular variable against time. This is called a time series. Examples could include...

- The noon temperature in your garden every Sunday
- The daily value of the FTSE 100
- Your height at yearly intervals
- A hardware shop's monthly sales

REMEMBER! In time series graphs, 'time' is always on the **x**-axis (horizontal).

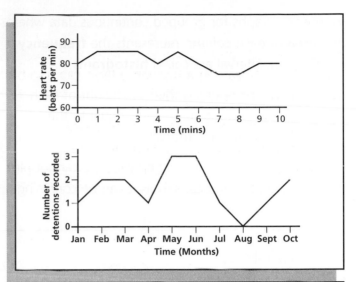

Seasonal Trends

Seasonality occurs when a cycle repeats itself. The first graph here is a time series showing the sales of Wellington boots in a High Street store. It is quite easy to find the period of the cycle by measuring peak to peak or trough to trough. Compare this to the graph of the FTSE 100 index over the same period. Clearly there is no seasonality at all here. In fact there is no obvious trend.

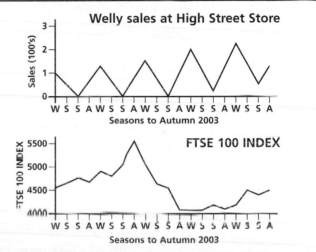

Moving Averages

Moving averages can be used to smooth out seasonal trends or even varying degrees of fluctuation. A three month moving average for welly sales at a store would be done in the following way:

Jan	Feb	Mar	Apr	May	Jun
145	181	169	106	100	142

The average for the first three months is found, then January drops out and is replaced by April and so on.

1st average $= \frac{145 + 181 + 169}{3} = \frac{495}{3} = 165$

2nd average $= \frac{181 + 169 + 106}{3} = \frac{456}{3} = 152$

3rd average $= \frac{169 + 106 + 100}{3} = \frac{375}{3} = 125$

4th average $= \frac{106 + 100 + 142}{3} = \frac{348}{3} = 116$

This example is a three point moving average because 3 pieces of data are used each time. Moving averages can make it easier to spot a trend in a set of data.

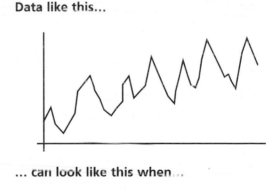

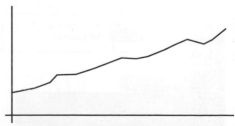

Data like this...

... can look like this when...

... smoothed out by taking a moving average.

Histograms

Frequency graphs for grouped continuous data where the **area** of each column represents the **frequency** of that class interval are called **histograms**.

With equal class intervals then the height of each column in our histogram is simply equal to the frequency of that class interval.

With unequal class intervals however, we cannot plot frequency on the vertical axis as it can be confusing.

We now plot a quantity called **frequency density**, which is related to the width of each class interval and the frequency of that class interval by…

Frequency Density (Height of column)	x	Width of Class Interval (Width of column)	=	Frequency of Class Interval (Area of column)

By plotting frequency density on the vertical axis then the area of each column represents the frequency of each class interval.

Example...

The frequency table alongside shows the distribution of the ages of the people in a supermarket at a particular time. Draw a histogram to illustrate this data.

Before we can draw our histogram we need to add a further column (in red) to our table for frequency density. It can easily be calculated by rearranging the relationship at the top of the page.

Frequency Density = $\dfrac{\text{Frequency of Class Interval}}{\text{Width of Class Interval}}$

We can now draw the histogram.

Age, A (years)	Frequency	Frequency density
$0 \leqslant T < 10$	10	$\frac{10}{10} = 1$
$10 \leqslant T < 20$	20	$\frac{20}{10} = 2$
$20 \leqslant T < 30$	60	$\frac{60}{10} = 6$
$30 \leqslant T < 40$	50	$\frac{50}{10} = 5$
$40 \leqslant T < 60$	60	$\frac{60}{20} = 3$
$60 \leqslant T < 100$	20	$\frac{20}{40} = 0.5$

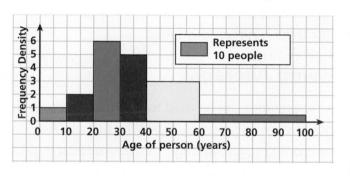

You can make sense of histograms with unequal class intervals if you remember that the area of each column represents the frequency occurring for that particular class interval.

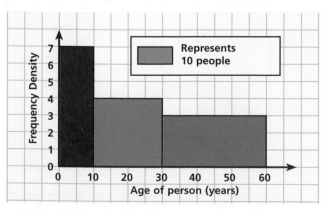

By just looking at the histogram it is not obvious what frequency (i.e. number of people) each column represents. However if we work out the area of each column.

Area of **A** is **7 x (10 – 0)** which represents 70 people.
Area of **B** is **4 x (30 – 10)** which represents 80 people.
Area of **C** is **3 x (60 – 30)** which represents 90 people.

Since we now know the value of the frequency occurring for each class interval we can analyse the distribution of the data shown in terms of modal class, median and mean.

A scatter diagram is a graph which has two sets of data plotted on it at the same time. When plotted the points may show a certain trend or correlation. A correlation is defined as the 'strength of relationship between two variables'.

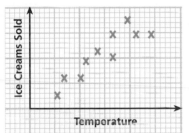

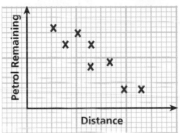

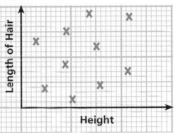

Positive Correlation

As one increases, the other also increases, e.g. number of ice creams sold and daytime temperature.

Negative Correlation

As one increases, the other decreases or vice-versa, e.g. amount of petrol left in tank and distance travelled by car.

Zero Correlation

No obvious trend between the two, e.g. length of hair and height. Remember that zero correlation does not necessarily imply 'no relationship' but merely 'no linear relationship'

Line of Best Fit

This is a straight line that passes through the points so that we have as many points above the line as we have below the line. A line of best fit can only be drawn if our points show positive or negative correlation. Opposite are four different examples of lines of best fit.

A and **B** are **POOR** lines of best fit. They both have the same number of points above and below the line but these points are bunched together and not spread out.

C and **D** are **GOOD** lines of best fit. They both have the same number of points above and below the lines and these points are spread out.

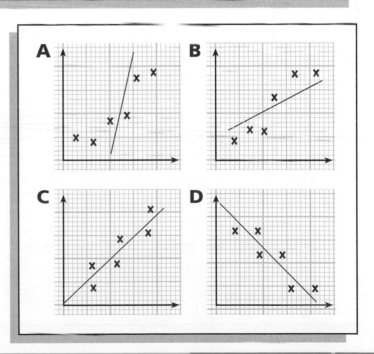

Scatter Diagrams 2

Example

Maths mark	60	30	46	89	48	78	26	38	91	20	65	73
Science mark	56	30	47	98	61	77	38	51	89	25	80	87

The table gives the maths mark and science mark for 12 pupils in their end of year examinations.

a) Draw a scatter diagram, including a line of best fit, to show the marks.

b) Tim was absent from his science exam but he achieved a mark of 65 in his maths exam. Use your graph to work out an estimated science mark for Tim.

c) Jenny achieved a mark of 42 in her science exam but she was absent for her maths exam. Again use your graph to work out an estimated maths mark for Jenny.

Solution

a)

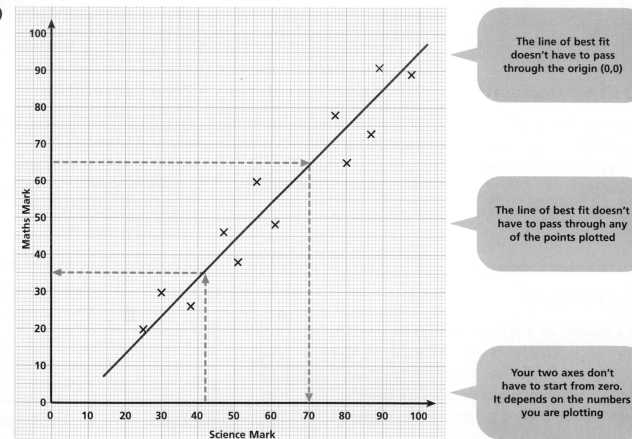

> The line of best fit doesn't have to pass through the origin (0,0)

> The line of best fit doesn't have to pass through any of the points plotted

> Your two axes don't have to start from zero. It depends on the numbers you are plotting

b) Go to 65 on the maths axis and then draw a dotted line ACROSS TO (→) the LINE OF BEST FIT and then DOWN TO (↓) the science axis. This is Tim's estimated science mark.

Answer is 70

c) Go to 42 on the science axis and then draw a dotted line UP TO (↑) the LINE OF BEST FIT and then ACROSS TO (←) the maths axis. This is Jenny's estimated maths mark.

Answer is 35

Drawing Pie Charts

Another way of displaying sorted data is to draw a pie chart. A pie chart is a circle which is split into different sectors. Here are the results of a survey carried out among 18 pupils to find their favourite sport.

Before we can draw our pie chart we need to calculate the ANGLE of the sector representing each sport. To do this we work out the fraction of the pupils for each sport, and then multiply this fraction by 360°.

We can now draw our pie chart. Always measure all angles carefully with a protractor and make sure that you always use the scale on your protractor that starts at 0°.

Favourite Sport	Frequency (no. of pupils)
Football	7
Tennis	3
Hockey	8
	Total = 18

Favourite Sport	Frequency of pupils	Angle to be drawn
Football	Total number of pupils → $\frac{7}{18}$ ← Number who liked football	$\frac{7}{18}$ x 360° = 140°
Tennis	Total number of pupils → $\frac{3}{18}$ ← Number who liked Tennis	$\frac{3}{18}$ x 360° = 60°
Hockey	Total number of pupils → $\frac{8}{18}$ ← Number who liked Hockey	$\frac{8}{18}$ x 360° = 160°
		Total = 360°

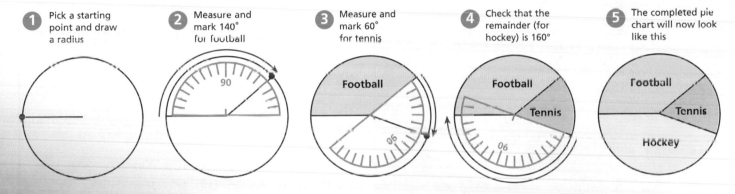

1. Pick a starting point and draw a radius

2. Measure and mark 140° for football

3. Measure and mark 60° for tennis

4. Check that the remainder (for hockey) is 160°

5. The completed pie chart will now look like this

Getting Information from Pie Charts

Getting information from a pie chart is very similar to drawing a pie chart except you need to work backwards. If the pie chart above shows the favourite sport for 18 pupils, the first thing we need to do is measure all the angles. We can then work out the number of pupils for each sport.

Favourite Sport	Fraction of circle	Number of pupils
Football	Total no. of ° for a circle → $\frac{140°}{360°}$ ← Number of ° for football	$\frac{140°}{360°}$ x 18 = 7
Tennis	Total no. of ° for a circle → $\frac{60°}{360°}$ ← Number of ° for tennis	$\frac{60°}{360°}$ x 18 = 3
Hockey	Total no. of ° for a circle → $\frac{160°}{360°}$ ← Number of ° for hockey	$\frac{160°}{360°}$ x 18 = 8
		Total = 18 pupils

Mean, Median, Mode & Range 1

Describing Data

We often use the word 'average' to describe a set of data and this is quite a useful tool. However, two batsmen who both average 35 in five innings could have had the following scores, showing quite different distributions:

Batsman A: 0, 10, 10, 60, 95

Batsman B: 30, 34, 36, 36, 39

In this case the 'average' score doesn't tell us the whole story. For this, we can use three different types of average, **mean**, **median** and **mode**, plus a description of the **range** of the data.

Example

Sam has the following coins in his pocket. What he has is a distribution of numbers (the different valued coins).

❶ Mean

This is the AVERAGE value and is given by:

$$\text{Mean} = \frac{\text{SUM OF ALL THE VALUES}}{\text{NUMBER OF VALUES}}$$

$$= \frac{1p+5p+1p+20p+10p+5p+1p+1p+10p}{9 \text{ (number of coins)}}$$

$$= \frac{54p}{9} = \mathbf{6p}$$

❷ Median

This is the MIDDLE value, providing that all the numbers have been arranged in order, lowest to highest.

1p, 1p, 1p, 1p, 5p, 5p, 10p, 10p, 20p ⟵ Arrange in order

1̶p̶, 1̶p̶, 1̶p̶, 1̶p̶, 5p, 5̶p̶, 1̶0̶p̶, 1̶0̶p̶, 2̶0̶p̶ ⟵ Tick off from the two ends to find the middle value

Median = 5p

If you tick off from the two ends and are left with two numbers in the middle, then the median is the number halfway between those two numbers (e.g. for 1̶,1̶, 1, 2,2̶,2̶ **Median = 1.5**)

❸ Mode

This is the MOST COMMON value, i.e. the number that occurs most frequently.

Mode = 1p   1p coin occurs more times than any other coin

❹ Range

This is the DIFFERENCE between the HIGHEST and the LOWEST value.

Range = 20p − 1p **= 19p**

Mean, Median, Mode & Range 2

Mean, Median, Mode and Range from a Frequency Table

Here is the frequency table (again) for the number of goals scored in games involving Germany, from Round 1 through to the semi-finals of the 2006 World Cup.

The data in this frequency table is discrete. The same processes would apply if the data was continuous.

Number of goals scored (x)	Frequency (f)	Frequency x no. of goals scored (fx)	
0	4	4 x 0 = 0	4 teams have scored 0 goals. Total number of goals scored = 4 x 0 = 0
1	1	1 x 1 = 1	1 team has scored 1 goal. Total number of goals scored = 1 x 1 = 1
2	4	4 x 2 = 8	4 teams have scored 2 goals. Total number of goals scored = 4 x 2 = 8
3	1	1 x 3 = 3	1 team has scored 3 goals. Total number of goals scored = 1 x 3 = 3
4	2	2 x 4 = 8	2 teams have scored 4 goals. Total number of goals scored = 2 x 4 = 8
	Total = 12	Total = 20	Total number of goals scored. = 0 + 1 + 8 + 3 + 8 = 20

Mean

To calculate the mean we need to add another column (in red) to our frequency table to calculate the total number of goals scored.

$$\text{Mean} = \frac{\text{Total number of goals scored (fx)}}{\text{Total frequency}}$$

$$= \frac{20}{12}$$

$$= 1.\dot{6} \text{ goals (per team per game)}$$

Mode

Mode = 0 and 2 goals

(because these occur more times than any of the others).

Median

Since we have 12 pieces of data, the median number of goals is halfway between the 6th and 7th piece of data (with an even number you always end up with two numbers in the middle).

4 teams scored	1 team scored	4 teams scored	1 team scored	2 teams scored
0 0 0 0	1	②②2 2	3	4 4

$$\text{Median} = \frac{2 + 2}{2}$$

$$= 2 \text{ goals}$$

Range

$$\text{Range} = 4 \text{ goals} - 0 \text{ goals}$$

$$= 4 \text{ goals}$$

Mean, Median, Mode & Range 3

Mean, Median and Mode from a Frequency Table Involving Grouped Data

Here is the frequency table for the recorded temperatures for various locations at home and abroad.

The data in this frequency table is continuous. The same processes would apply if the data was discrete.

Recorded temperatures, T (°C)	Frequency (f)	mid-temp. values (x)	Frequency x mid-temp. values (fx)
$10 \leqslant T < 15$	2	12.5	2 x 12.5 = 25
$15 \leqslant T < 20$	4	17.5	4 x 17.5 = 70
$20 \leqslant T < 25$	5	22.5	5 x 22.5 = 112.5
$25 \leqslant T < 30$	8	27.5	8 x 27.5 = 220
$30 \leqslant T < 35$	5	32.5	5 x 32.5 = 162.5
	Total = 24		Total = 590

These are class intervals

These are halfway values for our class intervals

Estimated Mean

With grouped data, the individual values are unknown. Therefore we have to use 'mid-temperature value' to provide an **estimate** of the mean. To calculate the mean this time we need to add two further columns (in red) to our frequency table.

$$\text{Mean} = \frac{\text{Total of recorded temperatures (fx)}}{\text{Total frequency}}$$

$$= \frac{590}{24}$$

$$= 24.58°C$$

Mode

Again we don't get an exact mode but we are able to determine which class interval or group is the MODAL CLASS. Modal Class is **25°C $\leqslant$ T < 30°C** since this class interval has the highest frequency (i.e. it occurs the most number of times).

Median

With continuous data we don't get an exact value for the median, but we are able to determine which class interval or group it is in. The above table has 24 pieces of data and so the median is halfway between the 12th and 13th piece of data. Using the frequency column, the median is in the **25°C $\leqslant$ T < 30°C** class interval.

Cumulative Frequency 1

Cumulative Frequency Diagrams

A cumulative frequency diagram is a useful way of displaying the distribution of a particular set of data.

Example

The grouped frequency table alongside shows the time taken by 80 girls to complete a piece of maths coursework. (The time was always rounded to a whole number).

Before we can draw the cumulative frequency diagram we need to add a further column (in red) to our grouped frequency table to show how this particular data 'builds up'.

All points are plotted at the UPPER BOUNDARY of each class interval, e.g. $0 < T \leqslant 10$, point is plotted at $T = 10$ etc.

Your curve always starts at the LOWER BOUNDARY of the first class interval, in this case $0 < T \leqslant 10$ so our curve starts at $T = 0$.

Your curve must be smooth and pass through all the points.

When you have pre-printed axes (as in the exam) only use the required section and check median, upper-quartile and lower-quartile carefully (see following page).

Time taken, T (minutes)	Frequency (i.e. no. of girls)	Cumulative Frequency
$0 < T \leqslant 10$	5	5
$10 < T \leqslant 20$	9	(9 + 5 =) 14
$20 < T \leqslant 30$	16	(16 + 14 =) 30
$30 < T \leqslant 40$	29	(29 + 30 =) 59
$40 < T \leqslant 50$	15	(15 + 59 =) 74
$50 < T \leqslant 60$	6	(6 + 74 =) 80

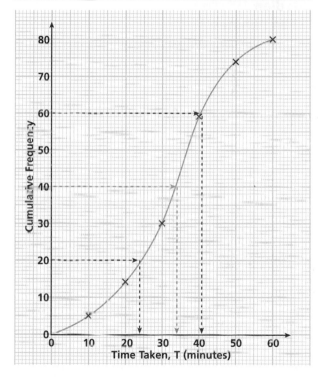

The Significance of the Shape

In the diagram alongside, curve A shows a very tight distribution whereas curve B shows a fairly broad spread. The tight distribution of curve A indicates very consistent data and is often associated with more reliable products, e.g. the durability of light bulbs, batteries, etc.

Using a cumulative frequency curve it is possible to read off the numbers of products (say light bulbs) which last less than a given amount of time. With a histogram you would have to perform a calculation.

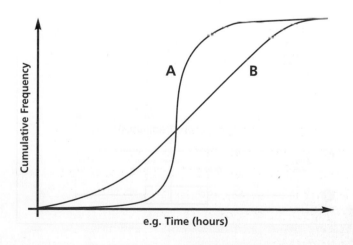

Cumulative Frequency 2

Median and Inter-Quartile Range

Once we have drawn the cumulative frequency diagram there are two ways we can compare the distribution of the data.

❶ Median

This is the middle value of our data. The median is obtained by drawing a line across (➤) from half way up the cumulative frequency axis (in our example this is at 40, i.e. $\frac{1}{2}$ of 80) to the curve and then down (▼) to give its value.

> **Median**
> **= 34 minutes**

❷ Inter-Quartile Range

This gives us a measure of the spread of the data about the median. Two values are needed.

i) A line is drawn across (➤) from three quarters of the way up the cumulative frequency axis (in our example this is at 60, i.e. $\frac{3}{4}$ of 80) and then down (▼). This value is called the upper quartile.

ii) A line is drawn across (➤) from one quarter of the way up the cumulative frequency axis (in our example this is at 20, i.e. $\frac{1}{4}$ of 80) and then down (▼). This value is called the lower quartile.

> **Inter-quartile range**
> **= Upper quartile - Lower quartile**
> **= 41 minutes - 24 minutes**
> **= 17 minutes**

Box Plots

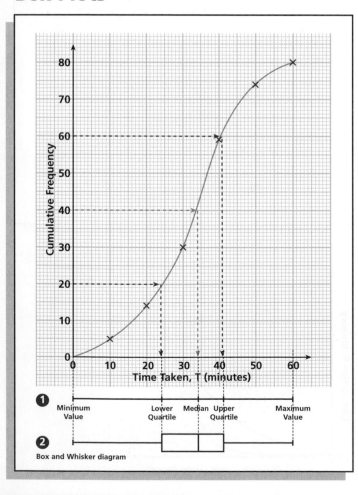

❶ Minimum Value — Lower Quartile — Median — Upper Quartile — Maximum Value

❷ Box and Whisker diagram

Box plots, or 'box and whisker' diagrams can show us how data is distributed in a pictorial form. It reveals the **MEDIAN**, the **QUARTILES** and the **RANGE**.

This can be quite useful, since most cumulative frequency graphs tend to have very similar shapes, which makes it quite difficult to compare one with another.

If we compare line ❶ at the bottom of our frequency curve with ❷, we can see that a box and whisker diagram presents this more visually by differentiating between the inter-quartile range and the full range.

This box and whisker plot shows a reasonably symmetrical distribution but diagram **(a)** below shows a positively skewed distribution with the median closer to the lower quartile while **(b)** shows a negative skew with the median closer to the upper quartile.

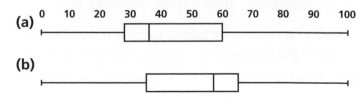

In your exam you may be asked to provide a **proof**. For example, to provide a proof that the sum of angles in a triangle is 180°.

A **mathematical proof** shows that a statement or expression is true in all cases.

For your exams, you need to be aware of both **algebraic** and **geometrical** proofs.

When answering a question which asks for something to be proved, always...

1 identify what you are being asked to prove – use a diagram if it helps

2 generalise the statement using symbols or letters

3 show the statement is true by using a logical set of arguments.

Here is an important point to remember:

An equals sign (=) shows that an equation is true for a certain value or values, e.g. $x(x + 1) = x^2 + 1$ is true for $x = 1$ but not for $x = 2$.

When it is true for all values, i.e. in an identity, then this symbol (≡) is used, e.g. $x + 2x ≡ 3x$ is true no matter what value x takes.

Geometrical Proof

Examples...

1 Prove that the opposite angles in this diagram are equal

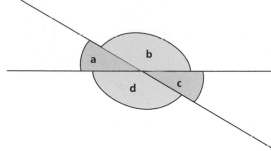

a + b = 180°
(because angles on a straight line add up to 180°)

a + d = 180°
(angles on a straight line)

So, **a + b = a + d**

Therefore, **b = d**
(because **b = 180° − a** and **d = 180° − a**)

Also, **a + b = 180°**
(angles on a straight line)

And, **b + c = 180°**
(angles on a straight line)

So, **a + b = b + c**

Therefore, **a = c**
(because **a = 180° − b** and **c = 180° − b**)

Proved!

Proof 2

2 Prove that the angle subtended by a semi circle is a right angle.

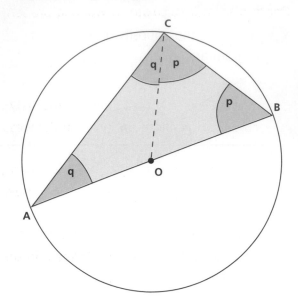

O is the centre of the circle

In △ OAC lengths OA and OC are equal (because OA and OC are both radii). Therefore, △ OAC is isosceles (because two sides are equal)

So, OÂC = OĈA (marked q)

Likewise, in △ OAC lengths OB and OC are equal (because OB and OC are both radii). Therefore, △ OBC is isosceles (because two sides are equal).

So, OB̂C = OĈB (marked p)

Using 'Angles inside a △ add up to 180°' (a corollary):

$$C\hat{A}B + A\hat{B}C + B\hat{C}A = 180°$$

So, $q + p + (q + p) = 180°$

$$2q + 2p = 180°$$

$$2(q + p) = 180°$$

Therefore $q + p = 90°$

Proved!

Algebraic Proofs

Examples...

1 Prove that the product of any two consecutive even numbers is always a multiple of 4.

Take two examples:
$2 \times 4 = 8$

remember, you multiply to find the product.

$34 \times 36 = 1224$

Write the numbers as **2n** and **2n + 2** (every even number can be written as another number multiplied by 2, so the first even number is 2n and then next consecutive even number is 2n + 2).

So, **2n(2n + 2)**   *Multiply out brackets...*

$$= 4n^2 + 4n$$ *... and simplify.*

$$= 4(n^2 + n)$$

Hence the product is always a multiple of 4.

2 Prove the identity:
$8(p - q) + 3(p + q) \equiv 2(p + 2q) + 9(p - q)$

You need to show that the left-hand side (LHS) equals the right-hand side (RHS):

LHS $\equiv 8(p - q) + 3(p + q)$ *Multiply out brackets and simplify.*

$$\equiv 8p - 8q + 3p + 3q$$

$$\equiv 11p - 5q$$

RHS $\equiv 2(p + 2q) + 9(p - q)$ *Multiply out brackets and simplify.*

$$\equiv 2p + 4q + 9p - 9q$$

$$\equiv 11p - 5q$$

Therefore, **LHS ≡ RHS**

3 Prove the identity:
$a^3 + b^3 \equiv (a + b)(a^2 - ab + b^2)$

RHS $\equiv (a + b)(a^2 - ab + b^2)$

$$\equiv a^3 - a^2b + ab^2 + a^2b - ab^2 + b^3$$

$$\equiv a^3 + b^3$$

Therefore, **LHS ≡ RHS**

Exam Preparation

The Day Before the Exam...
- Check that your calculator is working properly.

Before Entering the Exam...
- Check that your calculator screen is clear.
- Check that the MEMORY is clear.
- Ensure that DEG or D is showing so that any sin, cos or tan functions are given in degrees. It must not be showing RAD, R, GRA or G.
- Make sure that any FIX, SCI or ENG functions, which may affect your calculations, have been switched off.
- Many calculators have a RESET button on the back. It is a good idea to press this before the exam to restore the basic settings.

*Please note that some calculators may have different operating methods. Please check your own against the examples given.

Using a Scientific Calculator 2

Calculator Function	Example of use	Keys to press
SHIFT or **INV** or **2nd**	**SECOND FUNCTIONS** Press this button to use the 'second functions', which appear on the different buttons of your calculator above the main functions.	
x^2 SQUARE	$5^2 = 25$	5 x^2
$\sqrt{}$ SQUARE ROOT	$\sqrt{49} = 7$	49 $\sqrt{}$ or $\sqrt{}$ 49
x^3 CUBE	$4^3 = 64$	4 x^3 If you haven't got $\boxed{x^2}$ and $\boxed{\sqrt{x}}$ keys on your calculator, you can use the $\boxed{x^y}$, $\boxed{\sqrt[y]{x}}$ and $\boxed{x^{\frac{1}{y}}}$ keys to carry out the same functions.
$\sqrt[3]{x}$ CUBE ROOT	$\sqrt[3]{125} = 5$	125 $\sqrt[3]{x}$
x^y POWER	$4^6 = 4096$	4 x^y 6
$x^{\frac{1}{y}}$ or $\sqrt[y]{x}$ ROOT	$\sqrt[4]{6561} = 9$	6561 $x^{\frac{1}{y}}$ 4 or 4 $x^{\frac{1}{y}}$ 6561
$\frac{1}{x}$ or x^{-1} RECIPROCAL	$5^{-1} = \frac{1}{5} = 0.2$	5 $\frac{1}{x}$
$a^{\frac{b}{c}}$ FRACTION	To key in $2\frac{3}{5}$	2 $a^{\frac{b}{c}}$ 3 $a^{\frac{b}{c}}$ 5
$\frac{d}{c}$		Press $\boxed{\frac{d}{c}}$ to convert keyed-in mixed number into an improper fraction. Press $\boxed{a^{\frac{b}{c}}}$ again to convert improper fraction into a decimal.
[(... ...)] BRACKETS	$(5 + 2) \times 6 = 42$	[(... 5 + 2 ...)] × 6 =

Using a Scientific Calculator 3

Calculator Function	Example of use	Keys to press
sin **cos** **tan** **TRIGONOMETRY FUNCTIONS**	To calculate an unknown side length in a right-angled triangle x — 5cm — 30° $\sin 30° = \dfrac{x}{5}$ $x = 5 \times \sin 30°$ $= 2.5\text{cm}$	5 $\boxed{\times}$ 30 $\boxed{\sin}$ $\boxed{=}$ or 5 $\boxed{\times}$ $\boxed{\sin}$ 30 $\boxed{=}$
sin⁻¹ **cos⁻¹** **tan⁻¹** **INVERSE TRIGONOMETRY FUNCTIONS**	To calculate an unknown angle in a right-angled triangle 4.2cm 6cm x $\tan x = \dfrac{4.2}{6}$ $x = \tan^{-1}\left(\dfrac{4.2}{6}\right)$ $= 35.0°$ (to 1 d.p.)	4.2 $\boxed{\div}$ 6 $\boxed{=}$ $\boxed{\tan^{-1}}$ or $\boxed{\tan^{-1}}$ $\boxed{[(...}$ 4.2 $\boxed{\div}$ 6 $\boxed{...)]}$ $\boxed{=}$
EXP or **EE** **STANDARD FORM**	To key in 5.2×10^{6}	5.2 $\boxed{\text{EXP}}$ 6
Min **MR** **MEMORY**	Finally get used to using the memory buttons. $\boxed{\text{Min}}$ stores a number (usually the answer to a calculation) in the calculator's memory. $\boxed{\text{MR}}$ is used to recall the number when needed e.g. as part of a related calculation.	

Formulae Sheet

The Quadratic Equation:

The solutions of $ax^2 + bx + c = 0$, where $a \neq 0$, are given by:

$$x = \frac{-b \pm \sqrt{(b^2 - 4ac)}}{2a}$$

Volume of Prism = area of cross-section x length

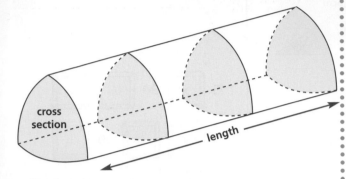

Volume of Cone = $\frac{1}{3}\pi r^2 h$

Curved Surface Area of Cone = $\pi r \ell$

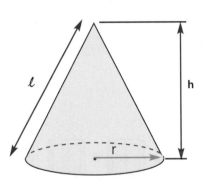

Volume of Sphere = $\frac{4}{3}\pi r^3$

Surface Area of Sphere = $4\pi r^2$

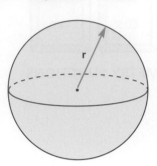

In any triangle ABC

Sine Rule: $\dfrac{a}{\sin A} = \dfrac{b}{\sin B} = \dfrac{c}{\sin C}$

Cosine Rule: $a^2 = b^2 + c^2 - 2bc \cos A$

Area of a Triangle: $\dfrac{1}{2}ab \sin C$

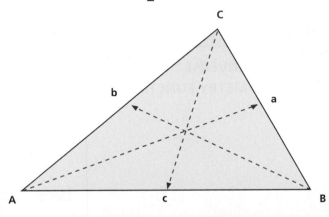

The formulae on the previous page are the ones that are included in the Edexcel specification. The additional formulae on this page are to help with your revision.

- nth term of an arithmetic sequence (where there is a common difference between terms).

  nth term = an + b

  where **a** is the common difference between the terms and **b** is an integer.

- The equation of a straight line graph

  y = mx + c

  where **m** is the gradient and **c** is the intercept.

- Pythagoras' Theorem (to find the length of an unknown side in a right-angled triangle).

  c² = a² + b²

  where **c** is the hypotenuse of a right-angled triangle, and **a** and **b** are the lengths of the other two sides.

- The trigonometric ratios (to find the length of an unknown side or the size of an unknown angle in a right-angled triangle):

  $$\sin\theta = \frac{opp}{hyp}$$

  $$\cos\theta = \frac{adj}{hyp}$$

  $$\tan\theta = \frac{opp}{adj}$$

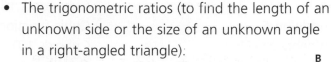

- In a circle...

  $$r = \frac{1}{2}d$$

  $$d = 2r$$

  circumference = 2πr or πd

  length of arc = $\frac{\theta}{360°}$ x 2πr

  area = πr²

  area of sector = $\frac{\theta}{360°}$ x πr²

  where **r** = radius, **d** = diameter and θ = angle at the centre of the circle.

- Volume of a cuboid

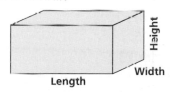

  volume = length x width x height

- Volume of a pyramid

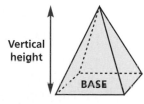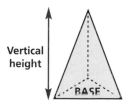

  volume = $\frac{1}{3}$ x area

  density = $\frac{mass}{volume}$

  speed = $\frac{distance}{time}$

Index 1

Acknowledgements

The author and publisher would like to thank everyone who has contributed images to this book:

All photographic images in this book are © 2006 Jupiterimages Corporation or © 2006 Lonsdale, a division of Huveaux Plc.

ISBN: 978-1-905129-78-2

Published by Letts and Lonsdale

Consultant Editor: John Proctor
Project Editor: Rebecca Skinner
Cover and Concept Design: Sarah Duxbury

This material has been endorsed by Edexcel and offers high quality support for the delivery of Edexcel qualifications.

Edexcel endorsement does not mean that this material is essential to achieve any Edexcel qualification, nor does it mean that this is the only suitable material available to support any Edexcel qualification. No endorsed material will be used verbatim in setting any Edexcel examination and any resource lists produced by Edexcel shall include this and other appropriate texts. While this material has been through an Edexcel quality assurance process, all responsibility for the content remains with the publisher.

Copies of official specifications for all Edexcel qualifications may be found on the Edexcel website - www.edexcel.org.uk

Notes

Notes